BASIC X-RAY SCATTERING FOR SOFT MATTER

Basic X-Ray Scattering for Soft Matter

Wim H. de Jeu

DWI – Leibniz Institute for Interactive Materials
Aachen, Germany

OXFORD

UNIVERSITY PRESS

UNIVERSITY PRESS

Great Clarendon Street, Oxford, OX2 6DP,
United Kingdom

Oxford University Press is a department of the University of Oxford.
It furthers the University's objective of excellence in research, scholarship,
and education by publishing worldwide. Oxford is a registered trade mark of
Oxford University Press in the UK and in certain other countries

Cover painting: 'Electromagnetic Charge' (1975)
by Alfred Jensen (1903–1981)

Cover painting reproduced by permission of the Estate of
Alfred Jensen (copyright VG Bild-Kunst, Bonn 2015).
Photograph by Nic Tenwiggenhorn courtesy of the owner of
the painting Museum Insel Hombroich (Neuss, Germany).

Epigram by Joël Dicker reproduced by permission of Editions De Fallois (Paris, France).

First Edition published in 2016

Published in the United States of America by Oxford University Press
198 Madison Avenue, New York, NY 10016, United States of America

British Library Cataloguing in Publication Data
Data available

Library of Congress Control Number: 2015957414

ISBN 978–0–19–872866–5 (hbk.)
ISBN 978–0–19–872867–2 (pbk.)

*To a new generation, in particular our sons
Jeppe, Maarten, Geert, and Bart*

Preface

This book grew out of lectures to first year graduate students of the Department of Polymer Science at the University of Massachusetts (Amherst, USA) during 2008–2010. On that occasion, I realized that the excellent textbooks on x-ray scattering I know and admire were mainly written by physicists, who used mathematics to make things clear. This was evidently not the approach that appealed to the students I was teaching. First, they had a rather variable background with often limited mathematics. Second, they were eager to know more about x-ray scattering as a user, but did not intend to become an expert. In my lectures I tried to accommodate these points. The notes were improved at various other occasions, notably lecturing in 2011 at the University of Science and Technology of China (Hefei, China) and in 2013 at the Leibniz Institute for Interactive Materials (DWI, Aachen, Germany).

Upon shaping the lecture notes into a book, I set several requirements. First, I aimed at a paperback of limited size that people would like to have on hand rather than on a shelf. Second, I wanted to include a large variety of examples of x-ray scattering of soft matter. Third, I liked, and so I kept, a chapter from my original notes on the different types of order/disorder in soft matter that play such an important role in modern self-assembling systems. Evidently, this combination did not make life easy, in particular regarding the coverage of the basics of x-ray scattering. I had great difficulties resisting inclusion of more material and it will be easy to point out subjects that are missing. Other people may feel there is still too much mathematics. Also, I did not aim at a practical guide to x-ray scattering but rather at explaining basic principles in a simple way. In spite of all the compromises, I believe there is a niche for a book of this type.

Necessarily, much material has been recycled from various sources. Usually this regards general knowledge that requires no specific reference. Still, I apologize to colleagues who might recognize a familiar sentence or certain aspects of a figure. I suggest considering it as a compliment to their efforts. However, I have tried to be complete in acknowledging the origin of experimental data. In fact, many of the examples are from my own work. Not so much because they are better than other ones, but because I happen to know them well, with all details readily available.

At the start of this project, the old book by D. W. L. Hukins, *X-ray Diffraction by Disordered and Ordered Systems* (Pergamon, 1981) was an important source of inspiration, especially in showing how valuable a small book can be. On a different level, I owe much to *Elements of Modern X-Ray Physics* by J. Als-Nielsen and D. McMorrow (Wiley, 2011). At some stage, I came to know the elegant book

Elementary Scattering Theory by D. S. Sivia (Oxford, 2011). It provides, in an approachable way, the mathematical toolbox for scattering that I did not want to emphasize. Hence, it is a perfect companion to this volume.

In the course of writing, I came across the work of Alfred Jensen. I learned that science provided the inspiration for much of his art and was a great part of the formulation of his diagrams and paintings. The Estate of Alfred Jensen kindly granted permission to reproduce one of his paintings on the cover of this book.

I am grateful to Björn Schulte and Hélène Freichels for their comments on the first complete version of the text, and to Khosrow Rahimi for his invaluable help in shaping the many figures. Finally, I want to express my gratitude to Martin Möller for providing, at DWI, the opportunity to continue my professional life after my formal retirement. At this stage of my career I am strongly motivated to share my experience in x-ray scattering of soft matter with a new generation of young researchers. I hope and trust that this book will find its way to their benefit.

Wim H. de Jeu
(www.wimdejeu.nl)

Contents

Un texte n'est jamais bon. Il y a simplement un moment ou il est moins mauvais qu'avant.

Joël Dicker, *La Vérité sur L'Affaire Harry Quebert*

Introduction and Overview

This first chapter is concerned with general aspects of x-rays. In the first section these will be positioned in a broad context. Subsequently, the generation of x-rays is discussed in terms of both conventional laboratory sources and modern synchrotrons. Next, the basics of x-ray scattering will be considered, first for a single electron, then for an x-ray beam incident on a composite sample of atoms or molecules. This is followed by a discussion of some general properties of x-rays as electromagnetic radiation: absorption, reflection, and coherence. The chapter finishes by comparing x-ray, neutron, and light scattering.

1.1 Introduction

X-rays were discovered more than a century ago, in 1895, by Wilhelm C. Röntgen. They almost immediately attracted much attention and curiosity among a wide public. This has generated many images of objects (see Figure 1.1a), of course all taken without appropriate protection. Especially after the Second World War, x-ray imaging had taken off for medical applications leading, these days, to highly sophisticated CT scans. This type of imaging basically depends on differences in absorption, and will not be discussed further here. In parallel, x-ray scattering or diffraction has developed into a valuable tool to investigate the structure of matter. In the first decades of the twentieth century, x-ray crystallography revealed the size of atoms, the length and types of chemical bonds, and the nature and differences of various materials on the atomic scale. These developments all relied on two prerequisites; first, knowledge of crystals and their basic properties which was essentially fulfilled at the time of Röntgen's discovery. The appearance and symmetry of crystals had fascinated scientists already for a long time, and in the course of the nineteenth century considerable knowledge and insight was built up. Second, it took some time and discussion to realize that x-rays were nothing other than electromagnetic radiation, just as visible light, but of a different (shorter) wavelength.

According to history, the idea that a crystal could be considered as diffraction grating (similar to an optical grating) came up in a discussion between Peter Ewald and Max von Laue in 1912. Ewald had developed a correct model of crystals but could not validate it with visible light. Von Laue realized that

Basic X-Ray Scattering for Soft Matter. Wim H. de Jeu.
© Wim H. de Jeu 2016. Published 2016 by Oxford University Press.

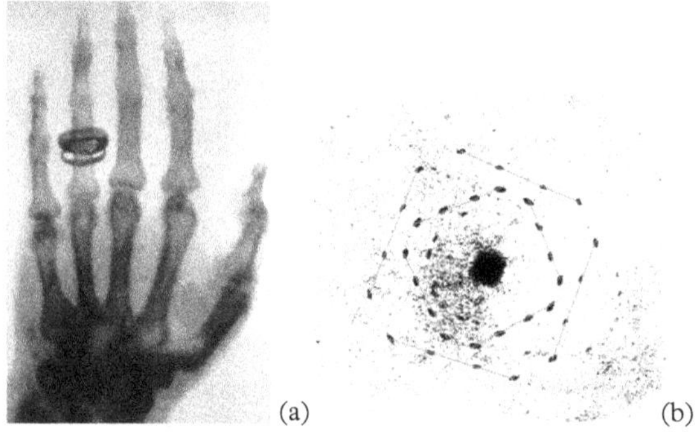
(a)   (b)

Figure 1.1 *(a) Original x-ray picture by Wilhelm Röntgen, indicating a difference in absorption. (b) X-rays pattern on a photographic plate after passing through a zinc blende (ZnS) crystal, as obtained by Max von Laue. The lines have been added to draw the eye to the symmetry.*

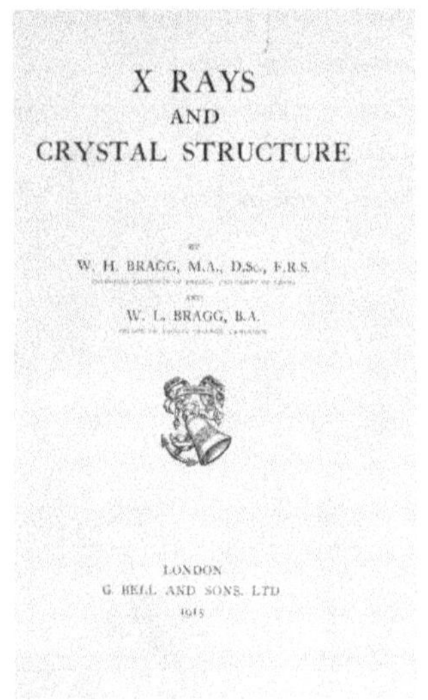

Figure 1.2 *Cover of the 1915 book by father and son Bragg.*

electromagnetic radiation of a shorter wavelength was needed to observe the small periods involved, and that x-rays might be the right choice. A subsequent experiment, in which a zinc blende single crystal was subjected to a beam of x-rays, produced a diffraction pattern on a photographic plate (Figure 1.1b). X-ray crystallography was born and von Laue received the Nobel prize in Physics in 1914.

Progress in x-ray crystallography was incredibly fast in those early years, most notably thanks to William Henry Bragg and his son William Lawrence Bragg. The latter developed the now well-known law that connects the scattering with reflections from equally spaced planes within a crystal. In the year of their common Nobel prize (1915), they had already published a book in the field (Figure 1.2). As early as 1917, Debye and Scherrer published their results on what nowadays is called powder diffraction. The understanding of the interaction between x-rays and matter increased steadily over the following decades. Important contributions to x-ray crystallography of biological molecules were due to Dorothy Hodgkin, who, in 1937, solved the structure of cholesterol and later penicillin and vitamin B_{12}, for which she was awarded the Nobel prize in Chemistry in 1964. In 1953, using x-ray fibre diffraction, Rosalind Franklin and Maurice Wilkins established the double helix structure of DNA.[1] Subsequently, the introduction of synchrotron sources in combination with increased computer power has given a considerable boost to the field. This has led to unravelling the structure and morphology of compounds of increasing complexity, including self-assembling molecular systems, large pharmaceutical compounds, and functional units of bio-organisms like proteins.

Most scientists get their first knowledge of x-ray scattering from elementary solid state textbooks that stress diffraction of crystal structures and the Bragg equation. Obviously this is the natural consequence of the developments in the field and the resulting great successes. However, it is somewhat unfortunate in connection with the field of soft matter, for which crystallization and crystal structure is not always the main issue. In this book, a separate chapter is devoted to order and disorder in soft matter, which goes well beyond simple crystalline order. The historical emphasis on crystals also has a semantic aspect. In the interaction of x-rays with matter, first the radiation is scattered by the electrons in atoms, molecules, particles, and crystals. Second, we have to consider the

---

[1] The road to the discovery of the double helix model of DNA (certainly one of the most important scientific milestones of the twentieth century) provides an intricate account of personal ambition and animosity in science. See, for example, Maddox, B. (2003) *Rosalind Franklin: The Dark Lady of DNA*. New York: Harper-Collins.

interference between the scattered waves, taking their phase difference into account (see Section 1.4). The combination of the two is called diffraction, leading to scattered intensity in specific directions where sufficient constructive interference occurs (x-ray 'peaks'). The expression 'diffraction' is used especially in the context of crystals. In a strict sense, 'scattering' refers only to the first part of the process. However, it is often used in a more loose way, even when some interference is also involved. Here we shall use scattering in this general way, like in the title of this book.

X-rays consist of electromagnetic radiation (typical wavelength $\lambda = 0.1\,$nm) that interacts with matter via the spatial distribution of the electrons. In contrast, neutrons are scattered by the atomic nuclei through nuclear forces. In electron diffraction (TEM or SEM) charged electrons interact with matter through Coulomb forces. This means that they feel the influence of both the positively charged atomic nuclei and the surrounding electrons. Because of these different forms of interaction, the three types of scattering are suitable for different types of study. TEM or SEM requires extensive effort to prepare thin samples. These methods give rise to beautiful pictures, that are, however, not always easy to quantify. X-ray and neutron scattering give directly quantitative structural information.

As scattering of x-rays depends on the electron distribution in the system, absence of differences in electron density obviously can only lead to unstructured diffuse scattering. Hence to get structural information, variation in electron density (contrast) is required. Starting at atoms, obviously the electron density varies proportional to the atomic number Z. For soft organic materials the variation of Z is relatively small.[2] Hence, in general, x-ray scattering from soft matter may suffer from limited contrast, which makes it a demanding field. On the other hand, the study of soft matter systems by x-rays is exciting and rewarding because of the wide variation in organization that can be observed, sometimes associated with important practical aspects and/or deep fundamental consequences.

1.2 Generation of x-rays

Like all electromagnetic radiation, x-rays can be considered either as waves or as photons. The photon energy is given by $E = h\nu = hc/\lambda$, where h is Planck's constant, c is the speed of light, and λ and ν are the wavelength and frequency, respectively.[3] Conventionally, x-rays are obtained by an electron bombardment of a water-cooled metal anode (often a copper target, see Figure 1.3a). The first effect is broad continuous spectrum of x-ray radiation produced by the (de)acceleration of the incident electrons when deflected by the charges inside the copper atoms.[4] The moving electrons lose kinetic energy, which is converted into photons because energy is conserved.[5] When the change of energy of the deaccelerated electrons increases, the resulting Bremsstrahlung becomes more intense while its maximum shifts toward higher frequencies (Figure 1.3b). However, in addition, some incoming electrons are absorbed while kicking out an

[2] C: $Z = 6$
N, CH: $Z = 7$
O, CH$_2$: $Z = 8$

[3] Especially at synchrotrons often the photon energy E is used instead of the wavelength λ, usually with E expressed in keV. The relation between the two then boils down to $\lambda = 1.24/E$ nm.

[4] This spectrum is generally known by its German name 'Bremsstrahlung'.

[5] A well-known example of charged particles undergoing acceleration is found when electrons move back and forth in an antennae producing electromagnetic radiation, such as transmitted by radio stations. In that situation, the dipole antenna serves as resonator, with standing waves of radio current flowing back and forth between its ends.

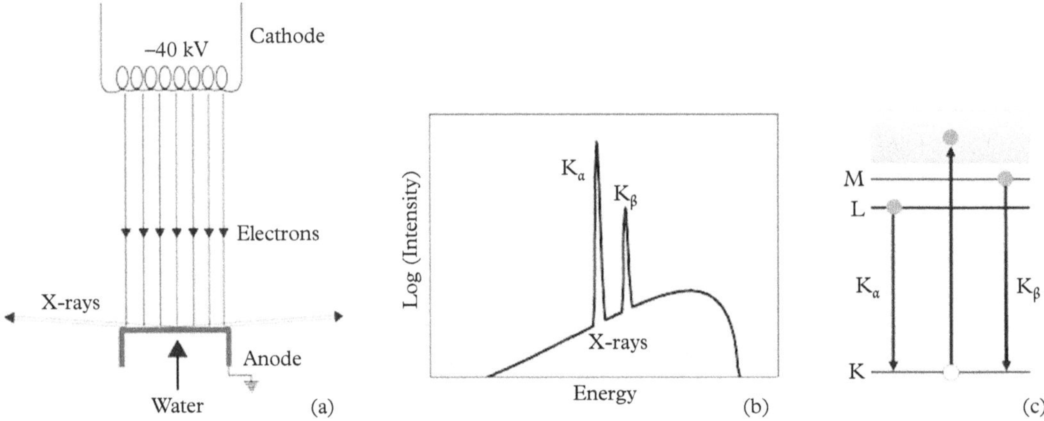

Figure 1.3 *(a) Standard x-ray tube. (b) The spectrum from the tube has discrete fluorescent lines superimposed on the continuous background. (c) Schematic atomic energy level diagram indicating transitions between an L and a K shell (K$_\alpha$ line) and between M and K (K$_\beta$ line).*

inner-shell electron into a higher energy level (Figure 1.3c). Upon relaxing to the original energy level, radiation is emitted.[6] The resulting CuK$_\alpha$ radiation has a wavelength $\lambda = 0.1542$ nm ($E = 8.05$ keV). In addition, an alternative transition gives CuK$_\beta$ radiation of a slightly smaller wavelength. The latter radiation is often removed by applying a Ni foil that absorbs K$_\beta$ radiation much stronger than K$_\alpha$ (see Figure 1.4a). At high resolution a fine structure is found (CuK$_{\alpha 1}$ and CuK$_{\alpha 2}$, see Figure 1.4b) which arises from a splitting of both the K- and the M-level into a doublet due to the electron spin. Some alternative target materials in use are Co ($\lambda = 0.071$ nm, 17.5 keV) and W ($\lambda = 0.021$ nm, 59 keV).

The power load that can be applied to a solid metal anode is restricted by the maximum heat dissipation for which the melting point sets a hard limit. Using rotating anode methods the maximum power can be increased by an order of magnitude. Rotating anode generators are complicated because a fast rotation is added

[6] The situation is analogous with the emission of optical radiation when an excited valence electron returns to its ground state (Bohr law). However, the inner-shell electrons involved here result in much shorter wavelengths.

Figure 1.4 *(a) The K$_\beta$ line can be suppressed using the absorption edge of Ni (broken line). (b) Due to the spin of the electron, each of the K-lines is split again: K$_{\alpha 1}$ and K$_{\alpha 2}$, K$_{\beta 1}$ and K$_{\beta 2}$.*

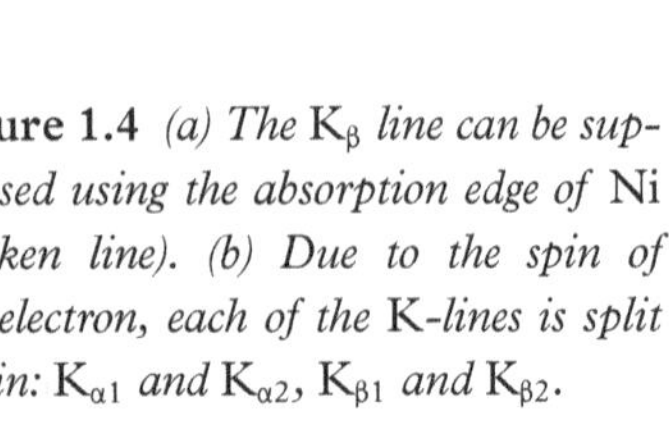

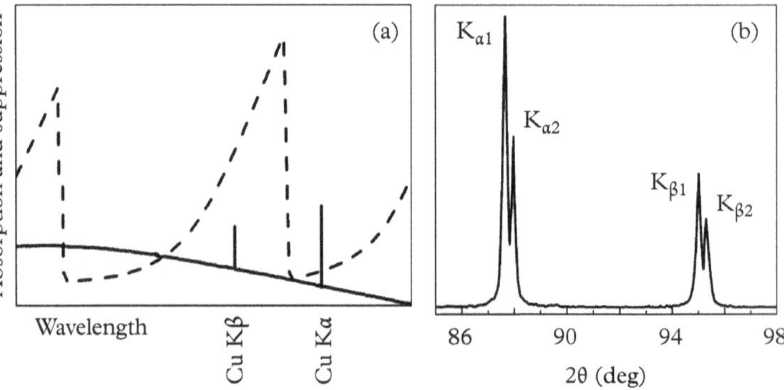

to a water-cooled anode at high voltage in vacuum. More recent developments use a micro-focused electron beam (spot size 5–50 μm) that can more easily be combined with advanced focusing x-ray optics. The latest progress involves an electron beam focused on a jet of liquid metal that serves as anode. Because of the large heat capacity and high anode speed, another order of magnitude more intensity can be generated, bridging the gap to synchrotron sources. Typically a Ga alloy is used providing GaK_α radiation of wavelength $\lambda = 0.134$ nm.

Synchrotron radiation consists essentially of Bremsstrahlung generated by electrons or positrons circulating in a storage ring (Figure 1.5). In fact, the ring is not circular but a multi-sided polygon. The electrons are accelerated at each of the bending magnets that are needed to keep them in their orbit (Figure 1.6). The relativistic speed multiplies the frequency of the circulating electrons with a large factor, into the x-ray range. Another dramatic effect of relativity is that the radiation pattern is strongly concentrated. When deflected from their straight path upon passing through a bending magnet, the electrons emit a rather narrow spray of x-rays tangentially to the plane of the electron beam (see Figure 1.6). The synchrotron light covers a wide and continuous spectrum, from microwaves to hard x-rays,

$$\sim\!0.3\,\text{nm} > \lambda > \sim\!0.02\,\text{nm} \quad \sim\!4\,\text{keV} < E < \sim\!80\,\text{keV}.$$

Apart from bending magnets, at third-generation synchrotron sources, like the ESRF (see Figure 1.7), radiation is emitted by so-called insertion devices in the

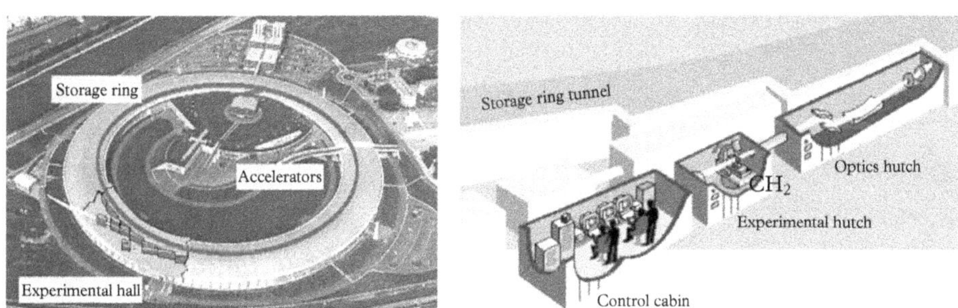

Figure 1.5 *(a) Schematic diagram of a storage ring (synchrotron) as used to generate x-rays. The ring is not circular but consists of multiple straight sections connected by bending magnets. (b) Typical layout of a beam line (Courtesy: ESRF).*

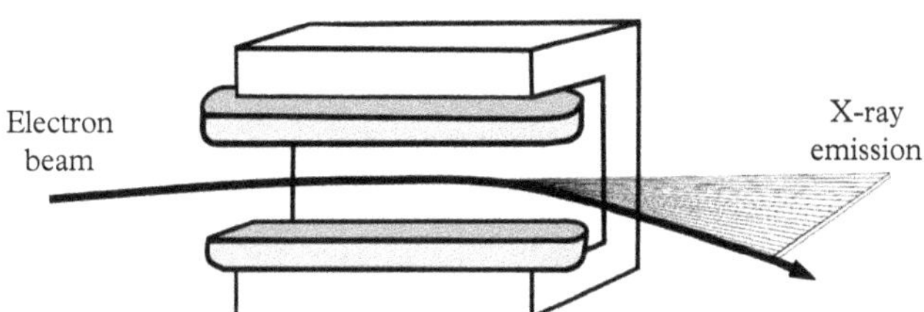

Figure 1.6 *Scheme of a bending magnet. As a result of the centripetal acceleration electromagnetic radiation is emitted.*

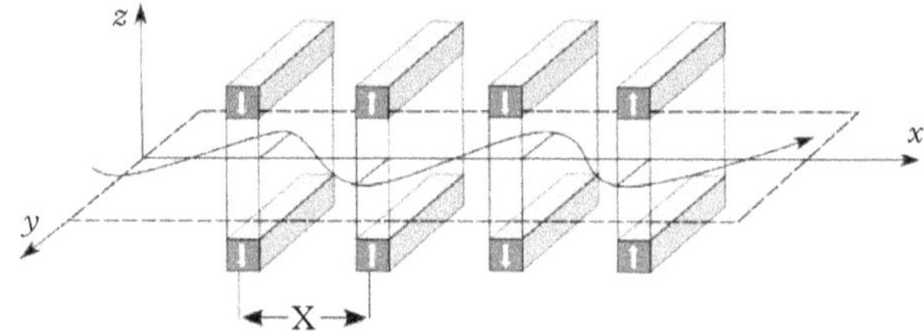

Figure 1.7 *Aerial view of the European Synchrotron Radiation Facility (ESRF) in Grenoble, France (Courtesy: ESRF). Note behind the ring the reactor of the Institut Laue–Langevin.*

Figure 1.8 *Principle of an undulator. The array of magnets with period X 'tunes' the coherent addition of radiation.*

straight sections of the ring. Strong fields from N alternating magnetic dipoles cause oscillations of the electron beam. The latter is, on average, not displaced. One distinguishes so-called 'wigglers' and 'undulators' that are physically very similar (see Figure 1.8). A wiggler can be considered a series of bending magnets concatenated together. Hence, the bandwidth remains broad and its radiation intensity scales with N. In an undulator the radiation is produced by the oscillating electrons, each interfering constructively, causing the radiation spectrum to have a relatively narrow bandwidth. The intensity of radiation scales as N^2.

The characteristics of synchrotron radiation are:

- Collimated emission.
- High intensity.
- Polarization.
- Continuous spectrum (except undulator).
- Beam size: $< 1\,\mu m$ to 1 mm.

Modern developments go in the direction of nano-beams of smaller sizes.

1.3 Scattering by a single electron

We consider an electromagnetic wave that hits a single electron. The wave is plane polarized: the electric vector vibrates along x in the xz-plane, in which z indicates the direction of propagation (Figure 1.9). Because of its electric charge, the electron (at the origin) starts vibrating along the x-axis, the direction of the incident electric field. As discussed already, accelerating charges generate electromagnetic radiation. Hence the electron will act as a source of x-rays itself, obviously with the same frequency (and thus the same wavelength) as the incident radiation. However, according to electro-magnetic theory, there will be a phase difference of π. If **E** is the electric field at the electron position, a force $-e\mathbf{E}$ will be experienced leading to an acceleration $e\mathbf{E}/m_e$ (m_e being the electron mass). The electron will act as a dipole antenna and radiate spherical waves with amplitude E_{rad} decaying as the inverse distance.

The acceleration of the electron leading to the scattered field can be calculated by classical electrodynamics. In terms of the scattered intensity $I_{\mathrm{rad}} = E_{\mathrm{rad}}^2$, the result is given by

$$I_{\mathrm{rad}}(R, t) = \left(\frac{e^2}{4\pi\varepsilon_0 mc^2}\right)^2 \frac{I_{\mathrm{in}}}{R^2}\cos^2\psi = \mathrm{r}_e^2 \frac{I_{\mathrm{in}}}{R^2}\cos^2\psi. \tag{1.1}$$

In this expression r_e is the Thomson scattering length,[7] also called classical electron radius. Using x-ray scattering, in expressions for $I(R,t)$ proportionality factors are often disregarded. Then any discussion is related to (differences in) relative intensities. The intensity of the scattered spherical wave depends on the distance R to the scattering centre and on the polarization direction ($\cos\psi$). At an observation point X in the original plane of polarization, the scattering will depend on ψ with no scattering left for $\psi = 90°$. In contrast, upon observation at X_1 in the zy-plane, there is no dependence on ψ_1 (see Figure 1.9). This situation

[7] $\mathrm{r}_e = e^2/(4\pi\varepsilon_0 mc^2) = 2.8179 \times 10^{-6}$ nm.

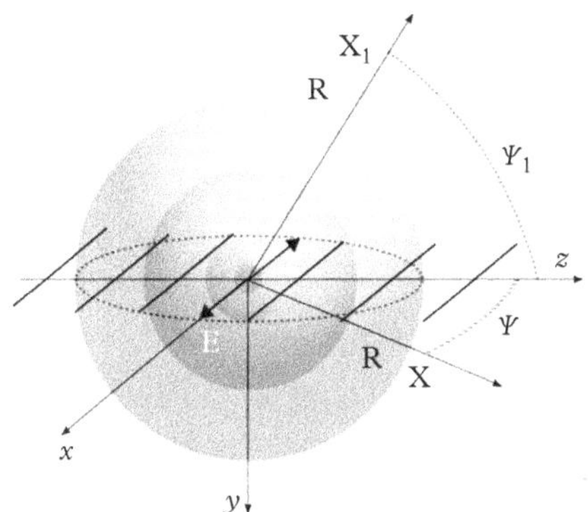

Figure 1.9 *Classical description of scattering of x-rays by an electron. The incident wave propagates along the z-axis while the electric field is polarized along x. The resulting scattering is different when the observation point lies in the polarization plane (X) or is normal to this plane (X₁) (after Als–Nielsen and McMorrow, 2011).*

applies to a synchrotron source that is polarized in the horizontal plane. In that case we find

$$P = 1 \qquad \text{for intensity scattered in the vertical plane}$$
$$P = \cos^2\psi \qquad \text{for intensity scattered in the horizontal plane.}$$

For an unpolarized laboratory source,

$$P = \left(1 + \cos^2\psi\right)/2.$$

In practice, any polarization factor is often added directly to the experiment as a correction to the scattered intensity. Note that the polarization correction can be disregarded for small angles where $\cos^2\psi \approx 1$. Evidently this is not the case anymore for scattering at larger angles.

1.4 Scattering of a plane wave

We now come to a more general description of scattering including interference and consider a plane wave front. The electric field of a sine-wave travelling along the z-direction (see Figure 1.10) can be described by $E_0 \cos\left[2\pi\left(\frac{z}{\lambda} - \frac{t}{T}\right)\right] = E_0\cos(kz - \omega t)$. Generalizing to an arbitrary direction defined by the wave vector $\mathbf{k}$, we obtain

$$E(r,t) = E_0\cos(\mathbf{k}\cdot\mathbf{r} - \omega t) = E_0\exp[i\,(\mathbf{k}\cdot\mathbf{r} - \omega t)]. \qquad (1.2)$$

[8] Note that $\exp(ix) = \cos x + i\sin x$ which allows writing a complex number as $z = a + ib = A(\cos x + i\sin x) = A\exp(ix)$. When using this form as in Equation (1.2) it is understood implicitly that we refer to the real part of the expression.

The term $\mathbf{k}\cdot\mathbf{r}$ gives the spatial dependence of the wave (a linear distance converted to a phase angle) and the ωt term gives the temporal dependence. Furthermore, we have generalized the expression using complex notation,[8] which allows easier manipulation.

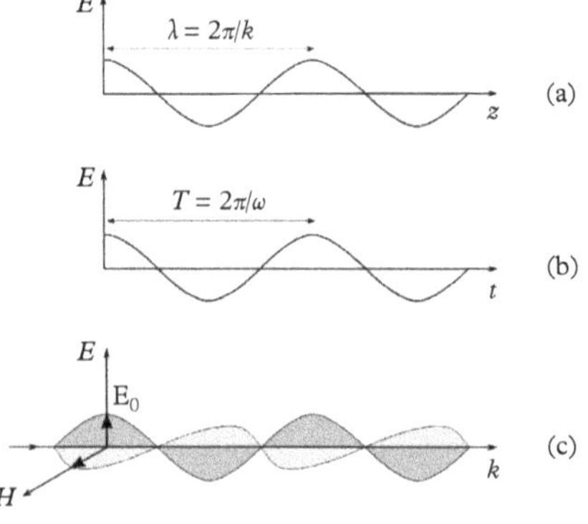

Figure 1.10 *Representations of an electromagnetic wave. (a) Spatial variation with wavelength* λ *or wave number* $k = 2\pi/\lambda$. *(b) Temporal variation with period* T *or angular frequency* $\omega = 2\pi\nu$. *(c) Travelling sine-wave along direction* **k** *with electric field* **E** *and magnetic field* **H**.

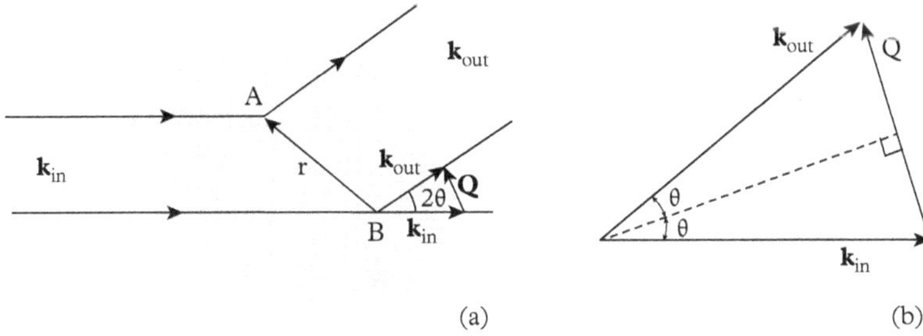

Figure 1.11 *(a) Representation of a typical scattering experiment. (b) Scattering triangle.*

When an x-ray photon interacts with matter, it is scattered by the electrons. This scattering is in principle elastic: no change of wavelength (classic) or equivalently no change of the energy $h\nu$ (quantum mechanics). Inelastic scattering (Compton scattering) involving energy transfer will not be considered here. For a typical elastic scattering experiment of an agglomerate of electrons (see Figure 1.11a) we can define (in the different notations that are in use)

$$\text{incoming beam:} \quad \mathbf{k} \equiv \mathbf{k}_{in} \equiv \mathbf{k}_i$$

$$\text{outgoing beam:} \quad \mathbf{k}' \equiv \mathbf{k}_{out} \equiv \mathbf{k}_f.$$

The scattering angle between $\mathbf{k}_{out}$ and $\mathbf{k}_{in}$ is given by 2θ. The scattered electric field is equal to a superposition of all secondary waves originating from the sample. Considering the scattering triangle in detail (Figure 1.11b), we note $\mathbf{k}_{in} + \mathbf{Q} - \mathbf{k}_{out} = 0$ defining the direction of $\mathbf{Q} = \mathbf{k}_{out} - \mathbf{k}_{in}$. Because $|\mathbf{k}_{in}| = |\mathbf{k}_{out}| = k = 2\pi/\lambda$ we easily derive

$$Q = |\mathbf{Q}| = 2k \sin\theta = \frac{4\pi}{\lambda}\sin\theta. \tag{1.3}$$

The vector $\mathbf{Q}$ is called scattering vector or wave vector transfer.[9] Just as a general wave vector $\mathbf{k}$ is related to the associated wavelength λ by $k = 2\pi/\lambda$, the scattering vector $\mathbf{Q}$ is related to a dimension d in the system by $Q = 2\pi/d$. Its properties will be further investigated in the next chapter (Section 2.1).

1.5 Absorption of x-rays

Obviously the intensity of x-rays passing through a sample decreases because of absorption. In a classical description, absorption of radiation is independent of the photon energy and can be described by (see Figure 1.12)

$$-dI = I(z)\mu dz,$$

$$\frac{dI}{I(z)} = -\mu dz.$$

[9] Sometimes the scattering vector is defined with a reversed sign: $\mathbf{Q} = \mathbf{k}_{in} - \mathbf{k}_{out}$, and/or writing $\mathbf{q}$ instead of $\mathbf{Q}$. In crystallography an alternative definition of the scattering vector is often used: $s = Q/(2\pi) = (2/\lambda)\sin\theta$.

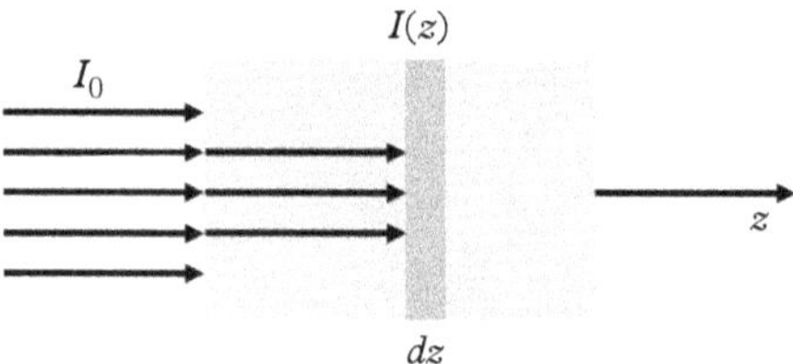

Figure 1.12 *Attenuation of an xray beam through a sample due to absorption (after Als-Nielsen and McMorrow, 2011). The intensity decreases exponentially with a characteristic coefficient μ.*

Defining I_0 as the incoming intensity at $z = 0$, the solution reads

$$I(z) = I_0 \exp(-\mu z). \tag{1.4}$$

Here, μ is the absorption coefficient given $\mu = (\rho_m N_A/M)\,\sigma_a$, in which ρ_m is the mass density, N_A Avogadro's number, M the molar mass, and σ_a the absorption cross section. $1/\mu$ is called the absorption length. At this distance the intensity has decreased by a factor $1/e$. Table 1.1 gives absorption values for some common materials, including polymers.

Practical x-ray experiments require an optimum sample thickness, compromising between maximum scattering (maximum volume $\rightarrow$ thick sample) and minimum absorption (thin sample). The optimum situation is reached for a sample thickness equal to the absorption length. From Table 1.1 we note that for hydrocarbons (including many polymers) a sample thickness of 1–2 mm can be expected to work well.

Quantum mechanically, when an x-ray photon is absorbed, the excess energy is transferred to an electron which is expelled from the atom (photoelectric absorption). If the hole is subsequently filled by an electron from an outer shell, radiation is emitted (x-ray–induced fluorescence). Absorption cross sections have a distinct dependence on the photon energy. Figure 1.13a shows the example of the noble gas krypton. Up to a photon energy of 14.32 keV the x-ray photon can only expel electrons from the L and M shells. The cross section varies approximately like $1/E^3$. At a characteristic energy—the so-called K-edge—the

Table 1.1 *Linear absorption coefficient μ for x-rays at $\lambda = 0.1542\,nm$ (CuK_α).*

	μ (mm^{-1})		μ (mm^{-1})
PS (C_8H_8)$_n$	0.4	silicon (Si)	14.1
PMMA ($C_5H_8O_2$)$_n$	0.7	beryllium (Be)	0.2
PVC (C_2H_3Cl)$_n$	8.6	nickel (Ni)	40.7
PBrS (C_8H_7Br)$_n$	9.7	lead (Pb)	256
quartz (SiO_2)	8.5	gold (Au)	417

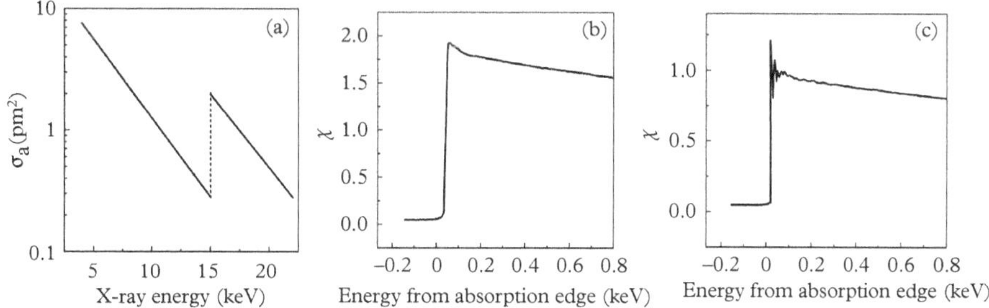

Figure 1.13 *(a) Absorption cross section of krypton with an edge above which a new 'channel' for absorption is opened up. (b) Structure around the edge for gaseous krypton. (c) Structure for krypton absorbed on graphite in which the atoms form a two-dimensional lattice (after Als-Nielsen and McMorrow, 2011).*
The function χ gives the fractional change in σ_a due to the presence of neighbouring atoms.

x-ray photons have enough energy to also expel a K-electron. A further treatment of these effects is beyond the scope of this book. However, we mention two consequences. First, the fine structure around an edge depends on the structure of the material. In Figure 1.13b and c this is illustrated for krypton as a gas and as a two-dimensional crystal absorbed on graphite. The wiggles around the edge demonstrate the phenomenon of x-ray absorption fine structure (XAFS). Second, using the tunability of the x-ray wavelength at a synchrotron source, one can vary the wavelength in the vicinity of an absorption edge of one of the constituent elements of the sample. This leads to a change in phase of the scattered x-rays that is unique from the rest of the atoms in the crystal.[10] Under these conditions, electrons are perturbed from their centrosymmetric distribution, which is reflected in a loss of symmetry in the pattern of scattered x-ray intensities. In a so-called MAD experiment,[11] sets of reflections are measured at three different wavelengths (far below, far above, and in the middle of the absorption edge). This allows solving the substructure of the anomalously diffracting atoms and hence the structure of the whole molecule.

1.6 Reflection and refraction of x-rays

Like all electromagnetic waves, x-rays show reflection and refraction at interfaces. These phenomena can be discussed assuming a homogeneous medium and the usual concept of refractive index. For x-rays we can write

$$n = 1 - \delta, \tag{1.5}$$

in which $\delta \approx 10^{-5}$ for many solid materials (compare with 10^{-8} for air). For a typical x-ray wavelength around 0.1 nm the associated frequencies, ν, are situated at the high side of any electronic resonances in solid material.[12] This permits us to

[10] This effect is called anomalous or resonance x-ray diffraction.

[11] MAD stands for Multiple wavelengths Anomalous Diffraction.

[12] $\nu = c/\lambda$ in which c is the velocity of light.

(a) Refraction and reflection

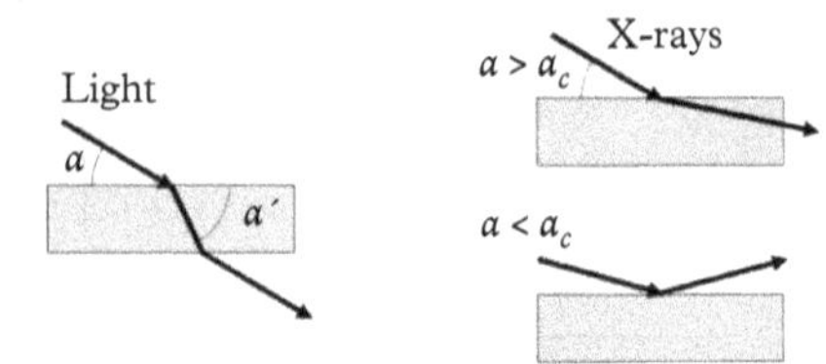

(b) Focusing X-ray mirror

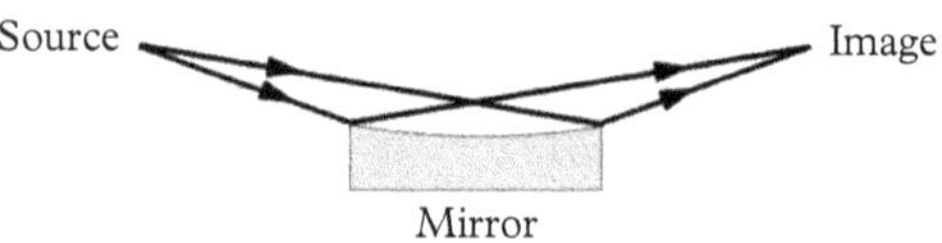

(c) Evanescent wave

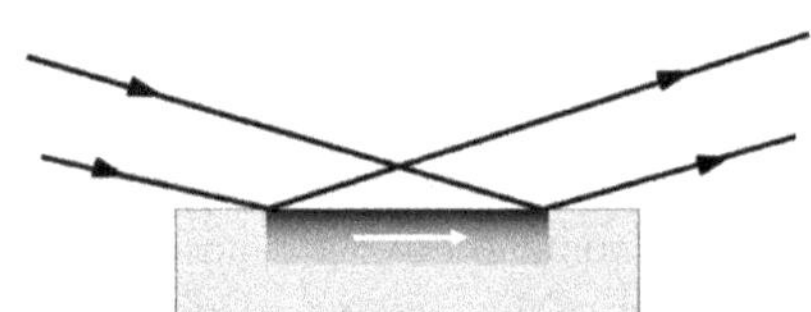

Figure 1.14 *(a) While for visible light the refractive index of glass is considerable larger than one, it is slightly smaller than one for x-rays. As a result total external reflection occurs at glancing angles below* $\alpha_c \approx 0.2°$. *(b) A focusing x-ray mirror can be constructed by a slightly bent surface and x-rays incident below the critical angle. (c) Below the critical angle x-rays penetrate typically about 10 nm in the material as an evanescent wave (after Als–Nielsen and McMorrow, 2011).*

calculate δ using a model of free electrons, resulting in $\delta > 1$ and a refractive index $n < 1$. As a consequence, x-rays show at an air interface total reflection below a critical angle $\alpha_c \approx 0.2°$ (Figure 1.14a). This can be contrasted with visible light for which n ranges typically from 1.3 to 1.8. In that situation a critical angle (Brewster angle) can only be observed going from an optical dense to a less dense material. Figure1.14b shows an important application of total reflection below the critical angle of x-rays in the form of a slightly curved focusing x-ray mirror. Figure 1.14c indicates the evanescent wave inside the material that is associated with total reflection. Its intensity decays exponentially along the surface normal, which gives a strong surface sensitivity. The evanescent wave can be considered as a 'new' incident x-ray beam in the plane of the surface to be used advantageously as a surface-sensitive x-ray probe (grazing-incidence diffraction; further discussed in Chapter 6).

Finally Figure 1.15 shows how in principle an x-ray lens can be constructed by cutting out the right shape in beryllium (a low-absorbing metal, see

Figure 1.15 *Comparison of (a) a converging lens for visible light with n > 1 and (b) a necessarily weakly converging one for x-rays with n < 1.*

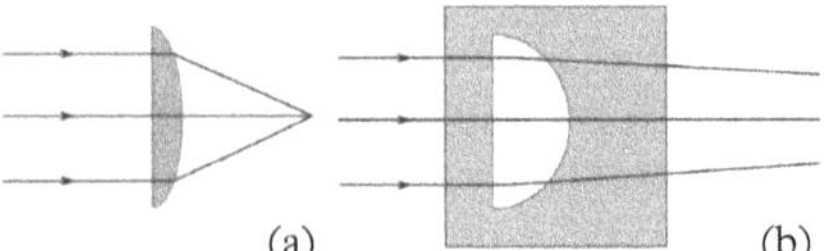

Table 1.1). In the form of a series of compound lenses (needed because of the weak convergence of individual ones) they are nowadays extensively in use at synchrotrons.

1.7 Coherent properties of x-rays

So far any incident x-ray beam has been considered as a perfect plane wave, which is by definition fully coherent. The phase varies uniformly along the direction of propagation. In reality an x-ray beam deviates from this picture in two ways:

- The beam is not perfectly monochromatic (dispersion).
- The beam is not perfectly parallel.

Each of these effects leads to a loss of phase information over a characteristic distance, the coherence length. The coherence length refers to the distance over which the phase information is fully lost (change of a factor π; exactly in antiphase). This is not different from electromagnetic waves in the optical region. Figure1.16a illustrates the effect of the wavelength dispersion, which determines the longitudinal correlation length ξ_l along the beam. At a synchrotron, the white beam is usually monochromatized by reflection against a 'perfect' crystal with a typical spread in wavelength $\Delta\lambda/\lambda \approx 10^{-5}$. For a wavelength of 0.1 nm the expression for ξ_l then leads to

$$\xi_l = \frac{0.1 \times 10^5}{2}\,\text{nm} = 5\,\mu\text{m}.$$

Figure 1.16b gives a similar calculation for the transverse coherence length. In this situation the non-parallelism of the beam is determined by the size of the source D and the distance from the source R. At a synchrotron, D is typically $20 \times 1000\,\mu\text{m}^2$ (V × H).[13] For a distance of, say, 40 m we then arrive at

$$\xi_{tV} = 100\,\mu\text{m}, \quad \xi_{tH} = 2\,\mu\text{m}.$$

These sizes are large enough for coherence to be used in experiments.

The coherence puts an upper limit on the separation of two objects if they are to give rise to interference. If we consider the scattering from two electrons separated by less than the appropriate coherence length, ξ, we have to add the amplitudes (taking the phase difference into account). Then we square to get the total scattered intensity. If they are further separated, they scatter independently. Then we square the two amplitudes separately and subsequently add them to get the total scattered intensity. This is not different from classical optics.

Recently, free-electron x-ray lasers have been built or are being built that provide intense pulses of coherent x-rays (Stanford, USA; Hamburg, Germany). In the meantime, coherent x-rays are being used at synchrotrons in spite of

[13] Vertical × horizontal.

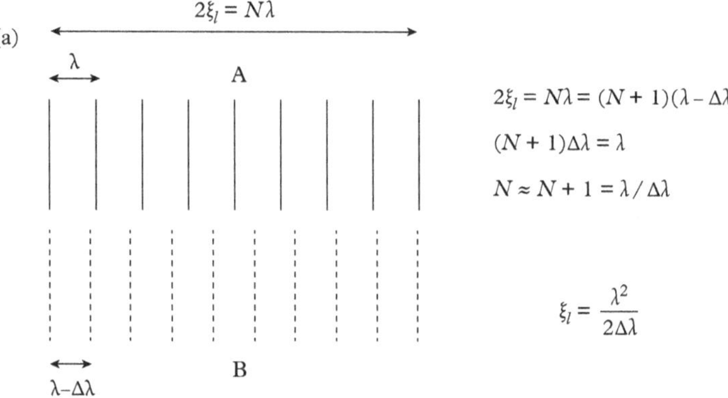

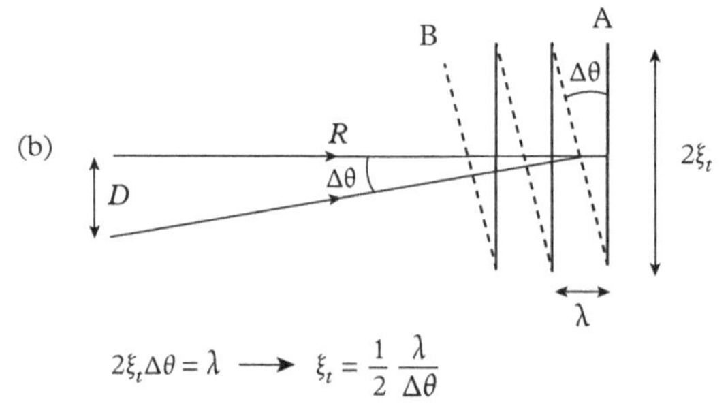

Figure 1.16 *Calculation of (a) the longitudinal coherence length ξ_l and (b) the transverse coherence length ξ_t (after Als–Nielsen and McMorrow, 2011).*
In (a) two waves A and B are compared that start with equal phases and differ by $\Delta\lambda$ in wavelength. After N respectively $N + 1$ periods along the beam they are again in phase. In (b) two waves are slightly non-parallel (difference in angle $\Delta\theta$) due to different origin positions at the source. After a transverse distance $2\xi_t$ they are again in phase.

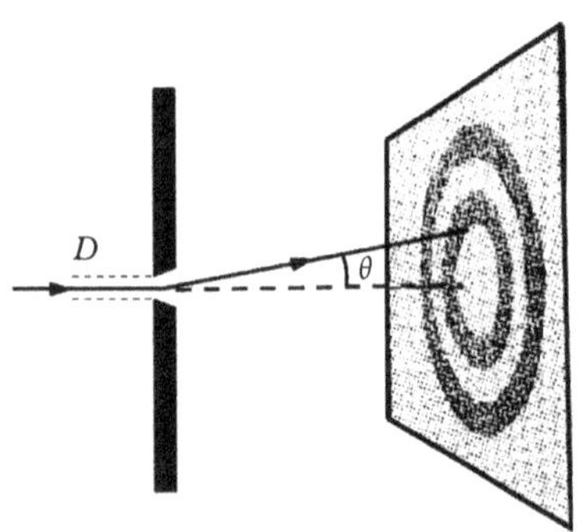

Figure 1.17 *Young's experiment leading to a coherent zero-order maximum.*

the small coherence lengths involved. To understand the principle, recall from classical high school optics Young's experiment in which light from a standard incoherent source falls on a pinhole. The well-known result on a far screen is a series of concentric maxima and minima (Figure 1.17). In this situation the zero-order maximum is by definition coherent. Similarly we can slit down the synchrotron beam by slits, leaving an opening smaller than the coherence lengths. A picture of the resulting x-ray beam after the slits is shown in Figure 1.18. The occurrence of a series of higher-order maxima and minima in two directions is very similar to that observed for a light source after a small rectangular hole. The zero-order maximum can be used as a secondary source of (partially) coherent x-rays. The drop in intensity compared to the primary beam is about a factor of 10^6.

A coherent beam of electromagnetic radiation—in standard optics from a laser—can be used to perform dynamic light scattering or correlation spectroscopy. It is well suited to probe the behaviour of complex fluids such as polymer solutions. Using coherent x-rays similar experiments can be done at a much smaller wavelength with a corresponding difference in penetration depth. In that case usually the term XPCS is used: x-ray photon correlation spectroscopy.

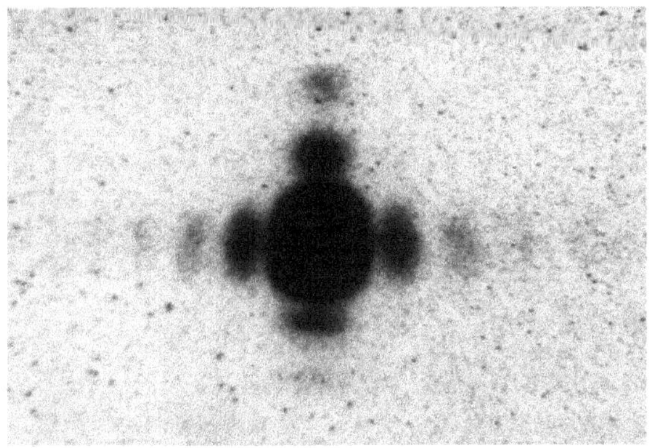

Figure 1.18 *X-ray intensity after a $0.72 \times 0.64\,\mu m^2$ slit at 35 m from a synchrotron source. (Reprinted with permission from Vlieg et al., 1997. Copyright International Union of Crystallography.)*

1.8 X-ray, neutron, and light scattering

Light scattering is a most convenient technique that can be carried out in any laboratory, thanks to the relatively small technical effort needed and the availability of intense coherent sources in the form of lasers. Obviously, the important difference with x-ray and neutron scattering is the much larger wavelength of typically 600 nm. Its effect is directly reflected in the accessible range of the scattering vector $Q = (4\pi/\lambda) \sin\theta$, which is inversely proportional to λ. This should be compared with $\lambda = 0.1542\,nm$, typical for CuK_α radiation from a laboratory x-ray source. Nevertheless, as shown in Figure 1.19, for small-angle scattering continuity between the two methods can be reached. Light scattering covers the Q-range from 0.008 to $0.04\,nm^{-1}$ (about 800 to160 nm) and x-ray scattering from $Q = 0.05\,nm^{-1}$ upwards (about 130 nm and smaller).[14] In fact, at synchrotron sources considerably smaller x-ray angles can be reached, leading to much larger overlap.

[14] Note that $1\,nm = 10\,\text{Å}$ and $1\,nm^{-1} = 0.1\,\text{Å}^{-1}$. To avoid the common confusion we shall adhere to the use of nm only and avoid Å.

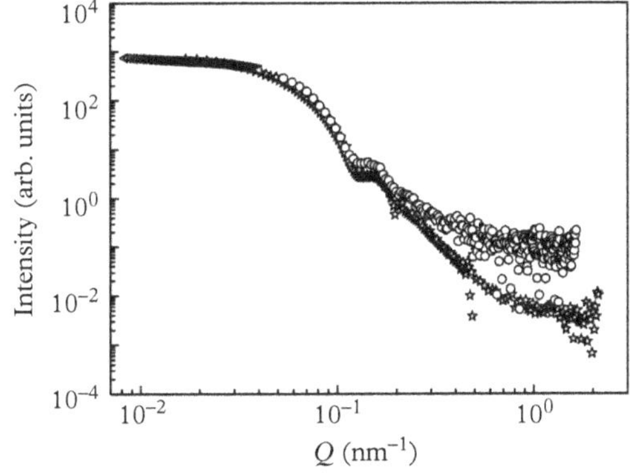

Figure 1.19 *Comparison of small-angle light scattering (triangles), small-angle x-ray scattering (circles), and small-angle neutron scattering (diamonds) for a concentrated solution of poly(N-isopropylacrylamide) microgel in water (Schmid, 2014).*

At large Q the x-ray and neutron results still agree within the experimental errors (in this region relatively large). The higher average level of the x-ray results probably indicates a problematic background subtraction.

Light scattering is restricted to optically transparent materials. Still multiple scattering is often a problem and the penetration depth in more turbid samples is small. As x-rays are scattered by the electrons surrounding the nucleus of an atom, heavy atoms scatter more efficiently than light ones. As a result, the penetration depth is smaller in the case of, for example, heavy metals. As noted in Section 1.1, this should not be a problem for soft matter for which (lack of) contrast is more of an issue. Nevertheless, if heavy atoms are incorporated, these may dominate the x-ray scattering pattern blurring other, smaller contrasts that may be present. In that context, neutrons come into the picture, especially for soft matter.

Neutrons scatter by interactions with the nuclei in the system. As the size of the scattering centres are many orders of magnitude smaller than the distance between the nuclei, neutrons can travel large distances in most materials. The scattering is determined by the neutron scattering length b that replaces the electron scattering length in the scattering formalism for x-rays. Note that b does not vary regularly with atomic number.[15] Most importantly, the b-values for ^{1}H and ^{2}H $\equiv$ D are very different. The negative sign for ^{1}H creates the remarkable possibility to make water as a solvent 'invisible' for neutrons by adjusting the ratio of normal water, H_2O, and heavy water, D_2O. Furthermore, by selective deuteration of specific positions in an organic material, a local contrast can be introduced. Such opportunities to play with hydrogen atoms are not possible for x-rays, to which they are effectively invisible. In addition, the magnetic moment of neutrons is non-zero, hence they are also scattered by magnetic materials. On the other hand, neutrons scatter weakly due to the limited flux from neutron sources. This makes neutron scattering a signal-limited technique. Hence neutron measurements may take days, while x-ray beam time is typically counted in shifts of eight hours.

The relevant wavelength for neutron scattering is given by $\lambda = h/(m_n v)$, in which h is Planck's constant and $m_n v$ the momentum of the neutron (m_n being the neutron mass and v its velocity). The wavelengths cover essentially the same range as used for x-rays (compare with Figure 1.19). The incident wave vector can be written as

$$(h/2\pi)\mathbf{k}_{in} = m_n \mathbf{v}. \tag{1.6}$$

Hence $\mathbf{k}_{in}$ is collinear with the velocity. Obviously $\mathbf{k}_{out}$ and the scattering vector can be defined as before: see Figure 1.11 and Equation (1.3).

While x-ray scattering can be done with laboratory sources as well as at a synchrotron, neutron scattering is restricted to large facilities.[16] These can be distinguished in (conventional) nuclear reactors and spallation sources. The neutrons as produced have energies in the range of MeV with corresponding wavelengths far too short to investigate condensed matter. Hence the neutrons are 'cooled down' by bringing them into thermal equilibrium with a 'moderating' material (water, liquid hydrogen) close to the source. The resulting energies down

[15] ^{1}H: $b = -3.74 \times 10^{-12}$ mm
^{2}H$\equiv$D: $b = 6.67 \times 10^{-12}$ mm
C: $b = 6.65 \times 10^{-12}$ mm
N: $b = 9.36 \times 10^{-12}$ mm
O: $b = 5.80 \times 10^{-12}$ mm

[16] Information on the relevant large facilities can be found at: http://www.lightsources.org and http://www.neutronsources.org.

to tens of meV take the wavelengths up to typical interatomic distances. A spallation source uses accelerators to make protons collide with a heavy metal target, driving neutrons from the nuclei of the target. The accelerators are coupled to a proton storage ring. As short proton bursts are produced, the resulting neutrons also arrive in pulses. A nice account of these processes and their applications is given by Pynn (1990).

It will be clear from the discussion so far that x-ray and neutron scattering have their own merits. X-ray scattering and crystallography have an enormous success story behind them in unravelling the structure of matter (compare with Section 1.1). This still continues today, mainly thanks to the versatility from laboratory anode sources to synchrotrons and free-electron lasers. The combination of coherence with high intensities is unique and is nowadays being extended from micro- to nano-beams. In comparison, neutron scattering suffers to some extent from the lack of a high-flux stream of neutrons. It is most valuable for special applications like deuterated and magnetic systems.

2

Basic Scattering by Particles

In Chapter 2 the scattering of x-rays is extended to atoms, molecules, and more complex particles like colloids. This allows the elucidation of the basic principles of scattering in some detail, taking the interference between scattering from electrons inside the scattering units into account. For this purpose, dilute solutions are looked at, for which interactions between the particles can be disregarded. This leads to the so-called *form factor* of the system, the modelling of which will be discussed in a case study. At the end of the chapter the *structure factor* will be introduced, the vehicle used to describe the interactions between particles.

2.1 Scattering by complex systems

2.1.1 Atoms, molecules, and crystal lattices

In Chapter 1 we discussed the scattering by a single electron (Section 1.3) and introduced the scattering vector $Q = |\mathbf{Q}| = (4\pi/\lambda)\sin\theta$ associated with an incident wave front. To describe scattering by more complex systems, obviously we have to introduce additional electrons. Let us start with two electrons in the scattering geometry used in Figure 2.1. An x-ray beam is incident along $\mathbf{k}_{\mathrm{in}}$ and scattered by a sample in the direction $\mathbf{k}_{\mathrm{out}}$ still with $k = 2\pi/\lambda$. The two electrons of the sample are situated at the origin B and at a point A with separation $\mathbf{r}$. Let the incident wave front with a uniform phase be along AA′,[1] after scattering along B′B. For the incoming beam the path difference between A′ and B equals $r\cos\alpha$, where α is the angle between $\mathbf{k}_{\mathrm{in}}$ and $\mathbf{r}$. The associated phase difference can be written as[2]

$$\frac{r\cos\alpha}{\lambda} \cdot 2\pi = kr\cos\alpha = -\mathbf{k}_{\mathrm{in}} \cdot \mathbf{r}.$$

Similarly, the phase difference between A and B′ is equal to $-\mathbf{k}_{\mathrm{out}} \cdot \mathbf{r}$. Hence the phase difference between waves from A and A′ at the resulting positions B′ and B is

$$\Delta\phi(\mathbf{r}) = -\mathbf{k}_{\mathrm{in}} \cdot \mathbf{r} + \mathbf{k}_{\mathrm{out}} \cdot \mathbf{r} = (\mathbf{k}_{\mathrm{out}} - \mathbf{k}_{\mathrm{in}}) \cdot \mathbf{r} = \mathbf{Q} \cdot \mathbf{r}.$$

[1] This situation is called far-field or Fraunhofer diffraction.

[2] Recall that the internal product of two vectors $\mathbf{a}$ an $\mathbf{b}$ is defined as $\mathbf{a} \cdot \mathbf{b} = ab\cos\alpha$, in which α is the angle between the two vectors.

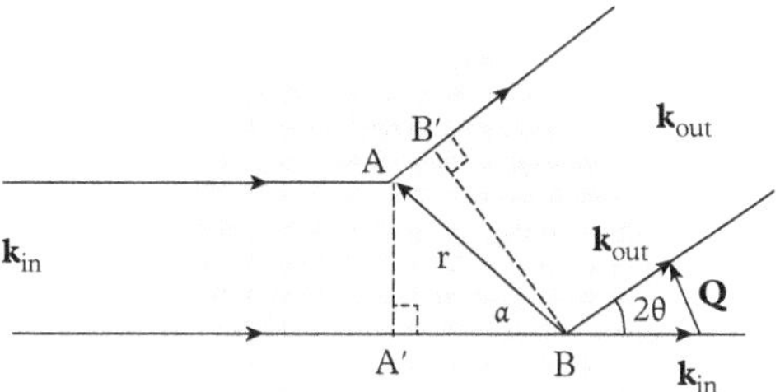

Figure 2.1 *Representation of a typical scattering experiment.*

The simplicity of writing a phase difference in terms of $\mathbf{Q}$ illustrates the usefulness of the concept of the scattering vector.

The reasoning so far can be easily extended to more scattering centres describing complex systems. Let the sample in Figure 2.1 represent an atom as in Figure 2.2a, and sum over all point scatterers (electrons). We consider the electrons in an atom as a spherical cloud around the nucleus. Making the transition to a continuum, the contribution from a volume element $d\mathbf{r}$ at position $\mathbf{r}$ with electron density $\rho(\mathbf{r})$ is proportional to $\rho(\mathbf{r})d\mathbf{r}$ with a phase factor $\exp(i\mathbf{Q}\cdot\mathbf{r})$.[3] Hence the scattering from an atom is determined by the so-called atomic scattering factor

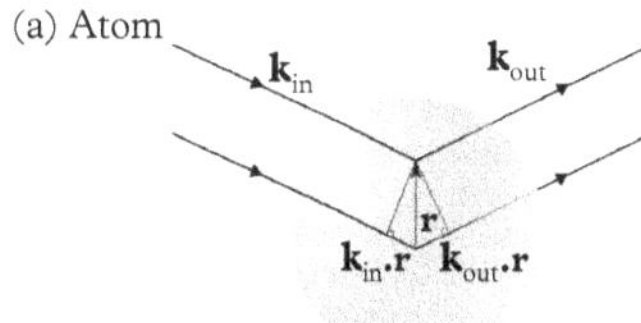

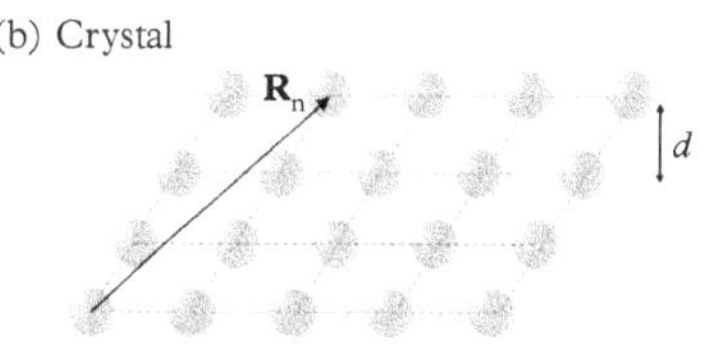

$$F^{\text{atom}}(\mathbf{Q}) = \int_{\text{atom}} \rho(\mathbf{r}) \exp(i\mathbf{Q}\cdot\mathbf{r})\,d\mathbf{r}. \tag{2.1}$$

Figure 2.2 *Scattering from an atom (a) and from a crystal lattice with spacing d (b) on which the atoms are organized with position vectors $\mathbf{R}_n$ (after Als-Nielsen and McMorrow, 2011).*

The scattered intensity $I(\mathbf{Q})$ of x-rays reaching a unit area of the detector in a certain time depends on the power of the beam (in a unit cross section). Hence the intensity is proportional to $|F(\mathbf{Q})|^2 = F(\mathbf{Q})F^*(\mathbf{Q})$, in which $F^*(\mathbf{Q})$ is the complex conjugate. Note that in the limit $Q \to 0$ all the different volume elements scatter in phase. Hence we find $F^{\text{atom}}(0) = Z$, the number of electrons in the atom. For increasing $\mathbf{Q}$ the different volume elements start to scatter out of phase and $F^{\text{atom}}(Q \to \infty) = 0$.

Equation (2.1) is quite general, and the summation can easily be extended to a combination of atoms forming a molecule (discussed in Section 2.3)

$$F^{\text{mol}}(\mathbf{Q}) = \sum_{\mathbf{r}_j}^{\text{molecule}} F_j^{\text{atom}}(\mathbf{Q}) \exp(i\mathbf{Q}\cdot\mathbf{r}_j). \tag{2.2}$$

Note that in Equations (2.1) and (2.2) we have left out the prefactor $-r_e$, needed to get the intensity in absolute units. Also, if the polarization factor differs sufficiently from unity, it is customary to correct the experimental data for this effect. If the scattered intensity can be measured for a sufficient number of $\mathbf{Q}$-values, one can (at least by trial and error) determine the positions $\mathbf{r}_j$ of the

[3] Compare with Equation (1.2).

atoms in a molecule. In practice a measurable signal requires that the molecules be assembled in a crystal.

A next step is to extend the summation over a crystal lattice (see Figure 2.2b). Let $\mathbf{R}_n$ be a vector defining the origin of a unit cell in a lattice and $\mathbf{r}_j$ that of the atoms in the unit cell. In general, a unit cell contains more atoms than the one pictured in Figure 2.2b. Then the position of any given atom in the sample can be described by $\mathbf{R}_n + \mathbf{r}_j$. In that situation we can generalize Equation (2.2) to

$$\begin{aligned}
F^{\mathrm{crystal}}(\mathbf{Q}) &= \sum_{j}^{\mathrm{unit\ cell}} F_j^{\mathrm{atom}}(\mathbf{Q}) \exp(i\mathbf{Q}\cdot\mathbf{r}_j) \sum_n \exp(i\mathbf{Q}\cdot\mathbf{R}_n) \\
&= F^{\mathrm{unit\ cell}}(\mathbf{Q}) \sum_n \exp(i\mathbf{Q}\cdot\mathbf{R}_n).
\end{aligned} \tag{2.3}$$

The first summation is the *unit cell structure factor* which involves a summation over all atoms in the unit cell. The second term is the *lattice sum* (discussed in Chapter 4).

As the formulas (2.1) to (2.3) are very basic to the scattering process, it is worthwhile discussing them a little further. In Section 2.1.2 we shall give a more geometrical discussion of scattering due to Ewald. Subsequently a few mathematical points connected to scattering as a Fourier transform will be summarized.

2.1.2 Scattering geometry and Ewald sphere

In Figure 2.3 we generalize the scattering triangle from Figure 2.1. The x-ray beam is incident at a specimen centred at O. In general the sample scatters x-rays in all directions, each of which corresponds to a vector $\mathbf{k}_{\mathrm{out}}$. The origin of the corresponding scattering vectors $\mathbf{Q}$ is at O′ at a distance $k = 2\pi/\lambda$ from O along the incident direction. If we complete the scattering triangle for each $\mathbf{Q}$, the end-points of $\mathbf{Q}$ lie on a sphere of radius k around O: the Ewald sphere. Hence for a fixed incident beam, a diffraction experiment probes values of $I(\mathbf{Q})$ for vectors $\mathbf{Q}$ with their end points on the spherical surface. To get actually scattered intensity while sampling the Ewald sphere, of course the condition for diffraction must be fulfilled. As we see, the Ewald sphere provides an elegant geometrical picture illustrating how scattering works.

As mentioned in Section 1.4, Q is related via $Q = 2\pi/d$ to a dimension d in the system. This can be illustrated in a somewhat different way using the Bragg equation $2d\sin\theta = \lambda$, from which we easily derive $\sin\theta = \lambda/(2d)$. Substituting this result in the definition $Q = (4\pi/\lambda)\sin\theta$, we arrive again at $Q = 2\pi/d$, the fundamental inverse relation between d (in real space) and Q (in reciprocal space or Q-space).[4] Note that large values of d give scattering at small scattering vectors Q (small scattering angles θ).

In a practical experiment, a flat-plate x-ray detector[5] is positioned perpendicular to $\mathbf{k}_{\mathrm{in}}$ at some distance behind the sample (Figure 2.4). The incident beam is assumed to be narrow and parallel. Obviously, only x-rays can be registered

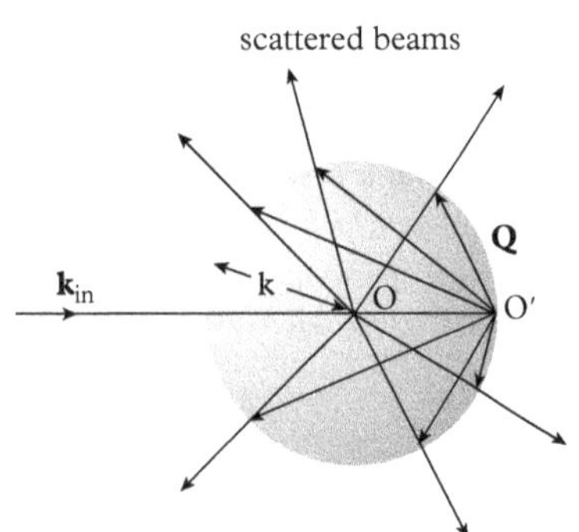

Figure 2.3 *Construction of the Ewald sphere.*

[4] The notion of reciprocal space is further developed in Chapter 4.

[5] In the old days a photographic plate was used, which was improved into a so-called image plate. Nowadays a whole series of electronic two-dimensional detectors is available (gas-filled wire detectors, charge coupled devices (CCDs), and pixel detectors).

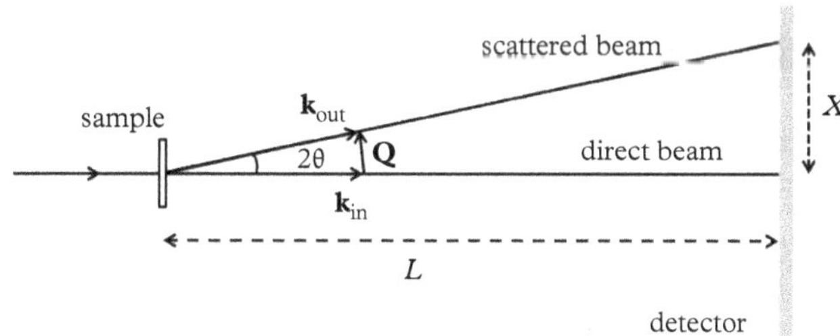

Figure 2.4 *Geometry of a general x-ray setup.*

that are scattered in the forward direction, which restricts the range of accessible Q-values (compare to Figure 2.3 with the sample at position O). The relation between the position of scattered intensity on the detector and Q is illustrated in Figure 2.4. Let a specific scattered beam be observed at a distance X from the transmitted one. Obviously X lies in the plane of the incident and the scattered beam, which—by definition—also applies to $\mathbf{Q}$. Using the sample-detector distance L, we can calculate the scattering angle from $\tan(2\theta) = X/L$. Knowing 2θ allows us to calculate Q and thus using $Q = 2\pi/d$ the corresponding period d.[6]

2.1.3   Scattering and Fourier transformation

Our basic Equation (2.1) for the scattering corresponds to the so-called Fourier transform of the electron density distribution $\rho(\mathbf{r})$:

$$F(\mathbf{Q}) = \int_V \rho(\mathbf{r}) \exp(i\mathbf{Q} \cdot \mathbf{r}) d\mathbf{r}, \tag{2.4}$$

in which V denotes integration over all space. The properties of Fourier transforms follow standard superposition: If two functions are added/rotated, the Fourier transformations also have to be added/rotated. As complex numbers are involved, the correct phase relation of the waves that are represented will be maintained. In particular, note that the following inverse Fourier transform is associated with Equation (2.4)

$$\rho(\mathbf{r}) = \int_V F(\mathbf{Q}) \exp(-i\mathbf{Q} \cdot \mathbf{r}) d\mathbf{r}. \tag{2.5}$$

Formally, a factor $(2\pi)^{-1}$ should be added to Equation (2.5) for each dimension involved. However, as we are not concerned with absolute x-ray intensity anyhow, we shall disregard prefactors and not include them. Both Equation (2.4) and (2.5) have the same properties and it does not matter which one is used first. The two integral forms differ only in the sign of their phase, which means that they are mutually inverted with respect to the origin.[7] The relation between electron density involved in x-ray scattering and conventional mass density is explained in Box 2.1.

[6] More formally we can write $2\theta = \tan^{-1}(X/L)$ and thus $Q = (4\pi/\lambda) \sin[\frac{1}{2}\tan^{-1}(X/L)]$.

[7] The same applies to a reversal of sign in the definition of $\mathbf{Q}$, as mentioned in Section 1.4.

Box 2.1. Relation between electron density and mass density

The electron density ρ of a specific material is related to the mass density ρ_m by multiplying the latter by Z_m, the number of electrons per unit mass. In formula:

$$\rho = Z_m \rho_m = N_A \frac{Z_M}{M} \rho_m,$$

in which N_A is Avogadro's number, Z_M the number of electrons per molecule or monomer unit of the material, and M the molecular mass. Further multiplication with r_e (see Section 1.3) leads to the scattering length density.

Equation (2.5) might suggest that we can directly obtain the density distribution of a system from the Fourier transform as measured by x-ray scattering. Unfortunately this is not true. X-ray scattering from a specific sample provides the intensity $I(Q) = F(Q)F^*(Q)$, from which we can determine the modulus or amplitude of $F(Q)$, but not its phase. Formulated differently, in general $F(Q)$ will be complex and from the measured $I(Q)$ we cannot get both its real and imaginary part. Both are needed in order to get the electron distribution by back-Fourier transforming $F(Q)$. The absence of phase information is called the 'phase problem' of x-ray diffraction analysis.

The physical reasons for the phase problem are as follows. First imagine a crystal with a perfect infinite periodic structure. This leaves the zero point of the periodicity (and thus the phase in a mathematical description) undefined and allows for a multitude of choices. Second, assume we could define the phase in such a periodic structure, for example by a single perfect surface by which the x-ray beam enters the sample. For x-rays of a wavelength of the order of 0.1 nm, the time period of the oscillations of the electric vector is of the order of 10^{-20} s. The intensity measurements from a typical detector will cover a large number of such oscillations. It will therefore be insensitive to when the waves arrive at various points at the surface and thus to the relative phases. The only way to solve this problem would be interference with a reference beam as is done in dynamic light scattering. This requires a coherent x-ray source like a laser, which is easily available for optical wavelengths, but not in the x-ray range.

A final point to be considered is the so-called convolution of two functions, say $\rho(\mathbf{r})$ and $\psi(\mathbf{r})$, defined by

$$\rho(\mathbf{r}) \otimes \psi(\mathbf{r}) = \int_V \rho(\mathbf{s})\psi(\mathbf{r}-\mathbf{s})d\mathbf{s}. \qquad (2.6)$$

Convolution of $\rho(\mathbf{r})$ means that it is smeared by the form of $\psi(\mathbf{r})$. The use of this concept can be simply demonstrated with a two-dimensional example. Figure 2.5 shows how a lattice of points can be convoluted by the shape of the

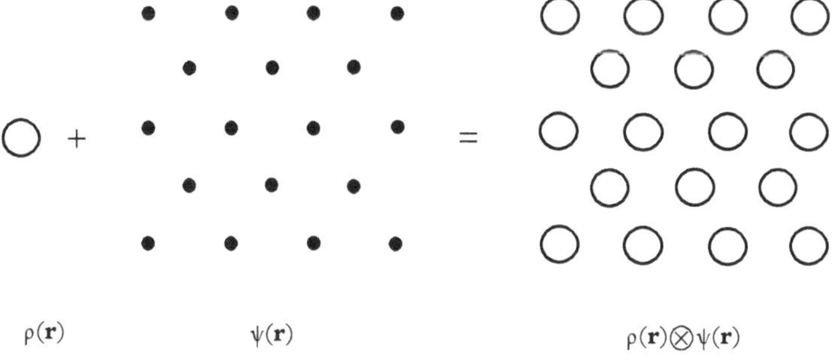

Figure 2.5 *Two-dimensional convolution of a function $\rho(\mathbf{r})$ with a set of points $\psi(\mathbf{r})$.*

constituting unit to give a real lattice.[8] With respect to x-ray scattering, convolutions have the following important property: The Fourier transform of a convolution of two functions is given by the product of the Fourier transforms of each of them. Hence, if the Fourier transforms of $\rho(\mathbf{r})$ and $\psi(\mathbf{r})$ are $F(\mathbf{Q})$ and $G(\mathbf{Q})$, respectively, we can write

$$\int \rho(\mathbf{r}) \otimes \psi(\mathbf{r}) \exp(i\mathbf{Q}.\mathbf{r})d\mathbf{r} = F(\mathbf{Q}) \cdot G(\mathbf{Q}).$$

A proof of this statement, for simplicity not in vector notation but as scalars, is given in Box 2.2.

Box 2.2. Calculation of the Fourier transform of a convolution

$$\int [\rho(x) \otimes \psi(x)] \exp(iQx)dx = \int \left[\int \rho(x_1) \cdot \psi(x-x_1)dx_1 \right] \exp(iQx)dx.$$

Substituting $y = x - x_1$ we can write

$$\iint [\rho(x_1) \cdot \psi(y)] \exp[iQ(x_1 + y)]\, dx_1 dy$$

$$= \int \rho(x_1) \exp(iQx_1)dx_1 \int \psi(y) \exp(iQy)dy$$

$$= F(Q) \cdot G(Q).$$

2.2 Diffraction of model scattering distributions

2.2.1 Diffraction of a 'top-hat' distribution

Let us return to Equation (2.1) describing the scattering from a spherical cloud of electrons. This situation applies to, for example, gases, liquids, and amorphous solids. To do so, we first consider a simple one-dimensional model:

[8] Crystal = motif $\otimes$ lattice.

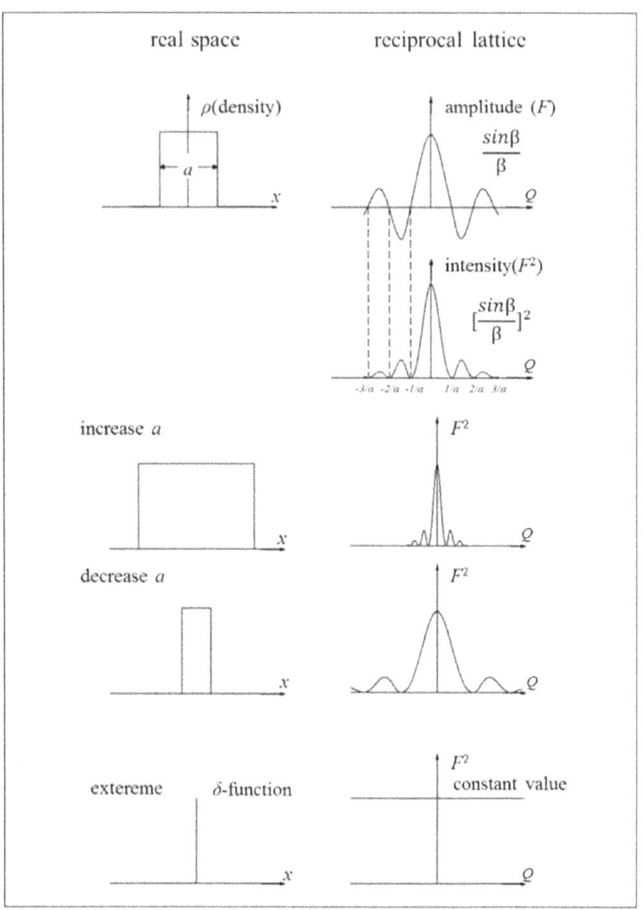

Figure 2.6 *Sketches (a to h) of the Fourier transform of a top-hat function upon varying its parameters ($\beta = \frac{1}{2}aQ$).*

the so-called top-hat function (Figure 2.6a). This function can be described by: $f(x) = 1$ for $-a/2 < x < a/2$ and $f(x) = 0$ otherwise. The Fourier transform then can be written as

$$F(Q) = \int_{-a/2}^{a/2} \exp(iQx)\,dx = \frac{1}{iQ}[\exp(iQx)]_{-a/2}^{a/2}$$

$$= \frac{i}{iQ}[\sin(Qx)]_{-a/2}^{a/2} = \frac{2}{Q}\sin\left(\frac{1}{2}aQ\right) = a\frac{\sin\left(\dfrac{1}{2}aQ\right)}{\dfrac{1}{2}aQ} \qquad (2.7)$$

The shape of the Fourier transform of the top-hat function is given in Figure 2.6b. Zero values occur at $\frac{1}{2}aQ = m\pi$ (m integer and $m \neq 0$) or $Q = 2\pi m/a$. Figure 2.6c–f indicate how $F(Q)$ and the intensity $F^2(Q)$ vary when the width of the top-hat function is varied. These results illustrate, once more, the inverse relation between real and reciprocal space. If a function in real space is less localized (broader top-hat), its Fourier transform in reciprocal space becomes more localized (narrower). In the opposite case of a smaller top-hat, the transform becomes broader and the zeros are further apart. Note the limit of a delta-function (Figure 2.6g: $a \to 0$, $\rho \to \infty$) which is perfectly localized, leading to a Fourier transform that is constant for all x (fully delocalized, Figure 2.6h). This inverse behaviour can be generalized to many other types of situation.

2.2.2   Diffraction from spherical objects

To discuss real particles, we expand the top-hat model to a spherical object in three-dimensional space. Suppose $\rho(r)$ depends only on $r = |\mathbf{r}|$ and is independent of its direction (spherical symmetric electron density). In that case, $F(\mathbf{Q})$ will again depend only on Q—the modulus of $\mathbf{Q}$—and not on its direction. Then we can expand Equation (2.7) to a spherical symmetric Fourier transform

$$F(Q) = \int_0^\infty 4\pi r^2 \rho(r) \langle \exp(i\mathbf{Q} \cdot \mathbf{r}) \rangle \, dr.$$

The brackets in $\langle \exp(i\mathbf{Q} \cdot \mathbf{r}) \rangle$ indicate a spherical average over all orientations. The calculation of this average is given in Box 2.3. Inserting the outcome in the integral, we arrive at

$$F(Q) = \int_0^\infty 4\pi r^2 \rho(r) \frac{\sin(Qr)}{Qr} \, dr. \tag{2.8}$$

The function $x^{-1} \sin x$ is plotted in Figure 2.7, and will be encountered in many scattering problems with spherical symmetry.[9] This symmetry can be either intrinsic or the result of spherical averaging (like in powder diffraction).

Box 2.3. Calculation of the average leading to Equation (2.8)

$$\langle \exp(i\mathbf{Q} \cdot \mathbf{r}) \rangle = \frac{1}{4\pi} \int_0^{2\pi} d\varphi \int_0^\pi \exp(iQr\cos\theta) \sin\theta \, d\theta$$

$$= \frac{1}{2} \int_0^\pi \exp(iQr\cos\theta) \sin\theta \, d\theta$$

$$= \frac{-1}{2iQr} \int_{iQr}^{-iQr} e^y \, dy = \frac{\sin(Qr)}{Qr},$$

in which the substitution $y = iQr\cos\theta$ has been made.

---

[9] This function is also written as $\mathrm{sinc}\, x = x^{-1} \sin x$.

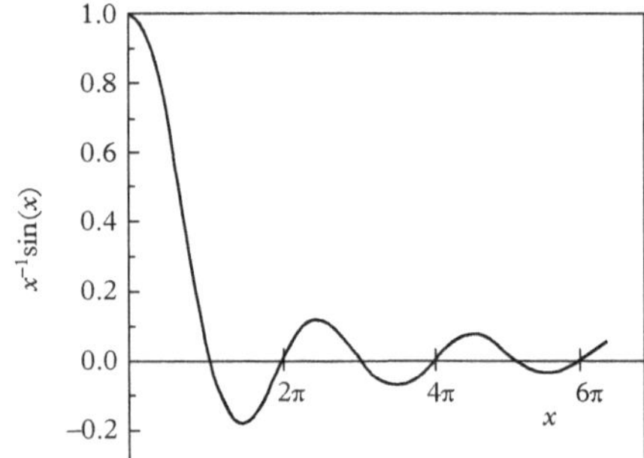

Figure 2.7 *The function $x^{-1}\sin x$.*

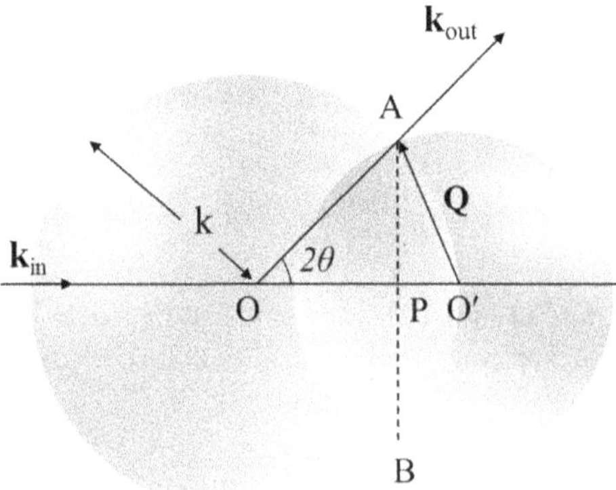

Figure 2.8 *Diffraction geometry for an isotropic scatterer.*

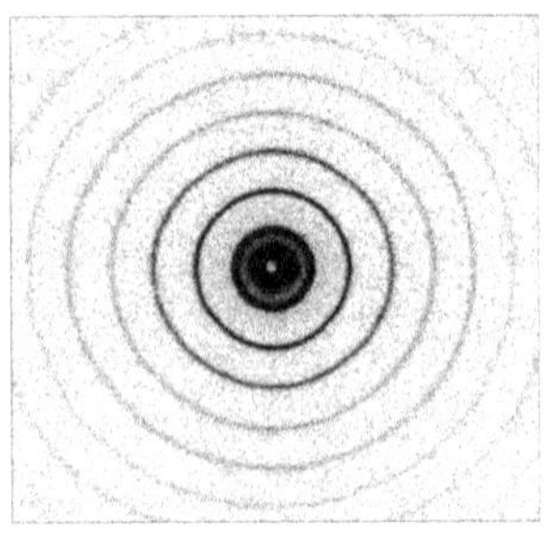

Figure 2.9 *X-ray diffraction pattern of a spherically symmetric sample.*

Let us return to a scattering distribution with spherical symmetry that scatters at a specific value of Q. This situation is depicted in Figure 2.8, in which the end points of Q lie on a spherical shell that intersects the Ewald circle at the points A and B. In three dimensions the Ewald circle becomes a sphere. Now A and B will lie on a circle whose centre is at P and which encloses a plane surface that is perpendicular to the plane of the figure. This circle forms the intersection with the Ewald sphere for which scattering can occur. According to Figure 2.4, such a distribution with spherical symmetry will show at the detector a diffraction pattern that has circular symmetry (concentric circles); see Figure 2.9.[10]

2.3   Atoms and molecules—Solutions

Atoms have spherical symmetry and $F^{\text{atom}}(Q)$ is essentially given by Equation (2.8). In principle, $\rho(r)$ can be calculated for an atom of each element, which is obviously a quantum-mechanical problem. Still, the above equation can be used to approximately calculate the scattering, because the scattering electrons are strongly bound. Figure 2.10 shows the atomic scattering factors for oxygen, carbon, and hydrogen. As $x \to 0$ we find that $x^{-1}\sin x \to 1$ and $F^{\text{atom}}(0) = Z$,

[10] A similar treatment, as given here for spheres, can be applied to the scattering by rods (fibres) eventually in combination with a helix. In those cases, the function $x^{-1}\sin x$ will be replaced by more complicated Bessel functions of various orders.

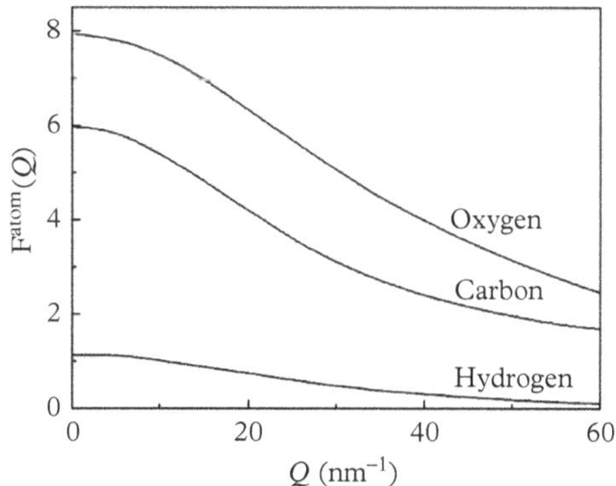

Figure 2.10 *Atomic scattering factors.*

the total number of electrons. The fall-off of $F^{\text{atom}}(Q)$ with increasing Q is due to interference of x-rays scattered in the same direction by electrons in different parts of the same atom. In contrast, for neutron scattering, $F^{\text{atom}}(Q)$ hardly changes with Q because the neutrons are scattered by the atomic nucleus which is very small compared to the wavelength. Hence interference effects between neutrons scattered by the same atom are negligible.

The above reasoning can easily be extended to a molecule, as given by Equation (2.2). Now the complex function $F^{\text{mol}}(\mathbf{Q})$ also allows—in addition to interference between scattering from electrons inside the atoms incorporated in $F^{\text{atom}}(\mathbf{Q})$—for interference effects between x-rays scattered by different atoms within the particular molecule (see Section 2.3.1). Interference between scattering from atoms in different molecules will be considered later. For $Q = 0$, the scattering still equals the total number of electrons, $F^{\text{mol}}(0) = \sum_j Z_j^{\text{atom}}$.

2.3.1 Particles at low resolution—Form factor

At small Q-values, $F^{\text{mol}}(Q)$ will be relatively insensitive to the details of the molecular structure. In different words: waves scattered at small values of Q correspond to a low resolution view of the structure. To explain the scattering in this regime it may be sufficient to consider the molecule as a simple shape with a uniform (average) electron density. Any orientation of the molecules will still be randomly distributed. Thus some globular macromolecules, such as certain enzymes, may be considered as spheres and certain polymers may be considered as rods in the analysis of the diffraction data.[11] Therefore, we shall discuss in some detail the case of x-ray scattering by spheres of uniform electron density. Applying Equation (2.8) to this situation we arrive at

$$F(Q) = \int_0^\infty 4\pi\, r^2 \rho(r)\, \frac{\sin(Qr)}{Qr}\, dr = \frac{4\pi\bar{\rho}}{Q^3} \int_0^a Qr\sin(Qr)\, d(Qr),$$

[11] The relevant Q-range is referred to as small-angle x-ray scattering (SAXS).

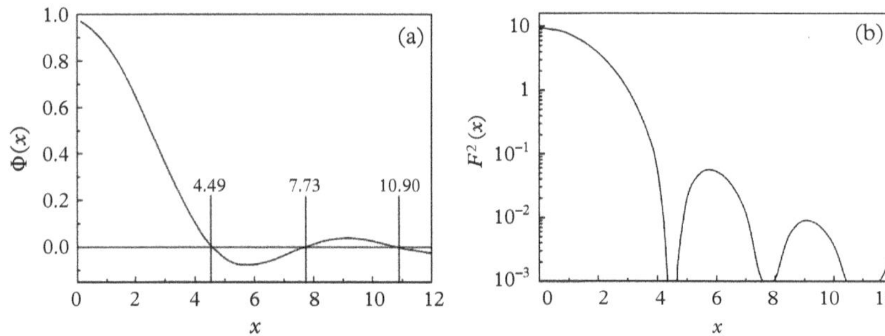

Figure 1.12

Figure 2.11 *(a) The form factor function $\Phi(x)$ of a sphere. (b) Corresponding values of $I(x) = F^2(x)$.*

in which the last equals sign refers to a sphere of radius a with average density $\bar{\rho}$. As $\int x \sin x\, dx = \sin x - x \cos x$, the result is

$$F(Q) = \frac{4\pi \bar{\rho}}{Q^3} \left[\sin(Qa) - Qa\cos(Qa)\right].$$

Writing $x = Qa$, we can introduce the *form factor function of a sphere*, given by[12]

$$\Phi(x) = 3(\sin x - x \cos x)/x^3, \tag{2.9}$$

and arrive at

$$F(Q) = (4\pi a^3 \bar{\rho}/3)\Phi(Qa) = n\Phi(Qa). \tag{2.10}$$

For $Q = 0$ we find once more that $F(Q)$ equals the total number of electrons, $n = 4\pi a^3 \bar{\rho}/3 = V\bar{\rho}$. Figure 2.11a shows the functional dependence of $\Phi(x)$ and Figure 2.11b shows the corresponding result for the intensity, $F^2(Q)$. The zeros of $\Phi(x)$ are obviously given by $\tan x = x$. Solving this equation is not trivial, as displayed in Figure 2.12. As before, we note that larger dimensions lead to more features at low Q. Hence the approximation of an average sphere becomes poorer at large values of Q (data at higher resolution).

The results of Equation (2.10) can be used to calculate the intensity at low Q-values for an ideal gas of molecules and for dilute solutions of colloids or (macro) molecules. In the latter case, we assume that the x-rays scattered by a solute particle do not interfere with x-rays scattered by the solvent or other solute molecules. Evidently, the last point requires that the solution is sufficiently dilute. Moreover, the background scattering from the solvent has to be subtracted. Obviously, a density contrast between solvent and dissolved particles is required. Hence, where we have written a density ρ in fact $\Delta\rho = \rho_{\text{solute}} - \rho_{\text{solvent}}$ is implied.[13] Using Equation (2.10) the resulting intensity from N spheres can be written as

$$I(Q) = NF^2(Q) = NV^2\bar{\rho}^2\Phi^2(Qa). \tag{2.11}$$

In Section 2.3.2 this equation will be considered in two limiting cases, leading to the Guinier- and Porod regimes at small and large Q-values, respectively.

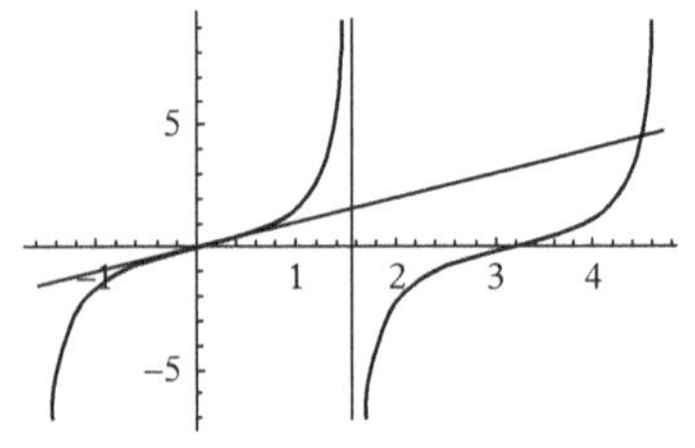

Figure 2.12 *Drawing* $\tan x$ *and* x *allows one to determine a first graphical solution of* $\tan x = x$ *at* $x = 4.493$. *The zero points are approximately given by* $(2k + 1)\pi/2$ *(k integer).*

[12] An alternative form is $\Phi(x) = (3/x)j_1(x)$, in which $j_1(x)$ is the spherical Bessel function of the first kind.

[13] Note that we cannot determine the sign of $\Delta\rho$. The scattering from 'white' particles in a 'black' sea is the same as from 'black' particles in a 'white' sea.

2.3.2 Guinier regime

The reasoning from the previous section has been extended to a method for deducing some information about the size of solute particles in ideal solutions via the Guinier law. Obviously, again, the validity is restricted to (macro) molecules at low concentrations and to low Q-values where the diffraction is insensitive to the details of the molecular structure. The Guinier law can be derived from Equation (2.11) by expanding $\Phi^2(Qa)$ from Equation (2.9)[14]

$$\Phi(x) \approx 1 - \frac{x^2}{10} + \dots \quad \text{or} \quad \Phi^2(x) \approx 1 - \frac{x^2}{5} + \dots$$

Using the latter result, Equation (2.11) can be written as the Guinier law in the form

$$I(Q) = NV^2\bar{\rho}^2\left[1 - \frac{1}{5}(Qa)^2 + \dots\right]. \tag{2.12}$$

When this result is compared to the more exact calculation of Equation (2.10), it turns out that the Guinier approximation works quite well.

Working in the Guinier regime, it is useful to introduce the radius of gyration R_G of the particles, which is a measure of the size of an object of arbitrary shape

$$R_G^2 = \frac{1}{V}\int_V r^2\,d\mathbf{r}.$$

From Table 2.1 we note that this formula reduces for spheres to $R_G = a\sqrt{3/5}$. In that case, Equation (2.12) takes the form

$$I(Q) = NV^2\bar{\rho}^2\left[1 - \frac{1}{3}(QR_G)^2 + \dots\right]. \tag{2.13}$$

In an alternative derivation, the Guinier law is written as

$$I(Q) = I_0\exp(-(QR_G)^2/3), \tag{2.14}$$

which is equivalent to Equation (2.13) if the expansion $\exp(-x) \approx 1-x$ is made.

The slope of a plot of $\ln[I(Q)]$ *vs.* Q^2 (Guinier plot) provides direct information about the size or radius of gyration of the scattering particles. In such a plot all intensity measurements have to be corrected for the solvent contribution, which can be measured in a separate experiment and subsequently subtracted. Note that knowledge of the radius of gyration is not sufficient to determine the actual shape because the interpretation relies on spherical averages.

[14] Recall $\sin x = x - \frac{x^3}{3!} + \frac{x^5}{5!} - \dots$ and $\cos x = 1 - \frac{x^2}{2!} + \frac{x^4}{4!} - \dots$

Table 2.1 *Radii of gyration of some simple bodies.*

Body	Defining dimensions	R_G
Sphere	Radius a	$a\sqrt{3/5}$
Large, thin disc	Thickness $2a$	$2a/\sqrt{2}$
Long, thin rod	Length $2l$	$2l/\sqrt{12}$
Random coil	Mean square end-to end distance $\langle h^2 \rangle$	$\sqrt{\langle h^2 \rangle/6}$

2.3.3 Porod regime

Figure 2.13 summarizes the various situations (to be) discussed. The so-called Porod regime refers to the large-Q limit of Equation (2.11). In that limit we make in the expression for $\Phi(x)$ (Equation 2.9) the approximation $\sin(Qa)/(Qa) \to 0$ (compare with Figure 2.7), leading to $\Phi(x) = (3/x^2)\cos x$. Substitution in Equation (2.11) gives

$$I(Q) \approx NV^2\bar{\rho}^2 \frac{9\langle\cos^2(Qa)\rangle}{(Qa)^4} \approx NV^2\bar{\rho}^2 \frac{9}{2Q^4 a^4}. \qquad (2.15)$$

To obtain the final form, the average value $\langle\cos^2(Qa)\rangle = 1/2$ has been inserted. Hence, for a sphere, the fall-off of the intensity with Q at higher angles should follow a Q^{-4} dependence (Porod law). This behaviour can be conveniently checked via a Porod plot of $Q^4 I(Q)$ *vs.* Q. In Equation (2.15) the factor V^2/a^4 relates the surface area to the unit volume. Hence it varies with the shape of the particles (spherical vs. cylindrical or disc-like) as well as with their smoothness. Hence for more complex shapes and and/or rough surfaces of the scattering entities the slope at high Q will not equal −4 anymore.

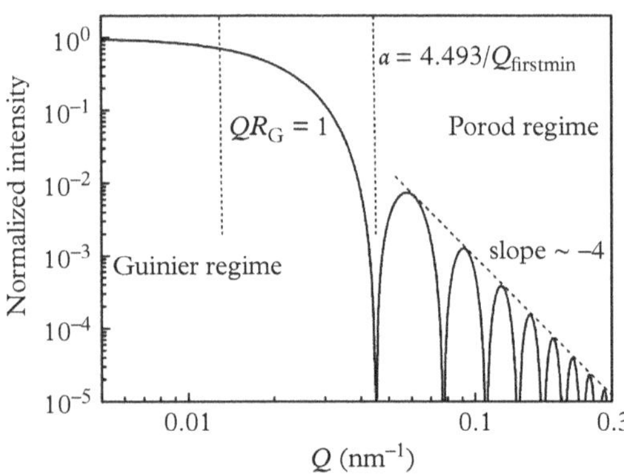

Figure 2.13 *Calculated SAXS pattern of non-interacting uniform colloid spheres with radius a = 100 nm.*

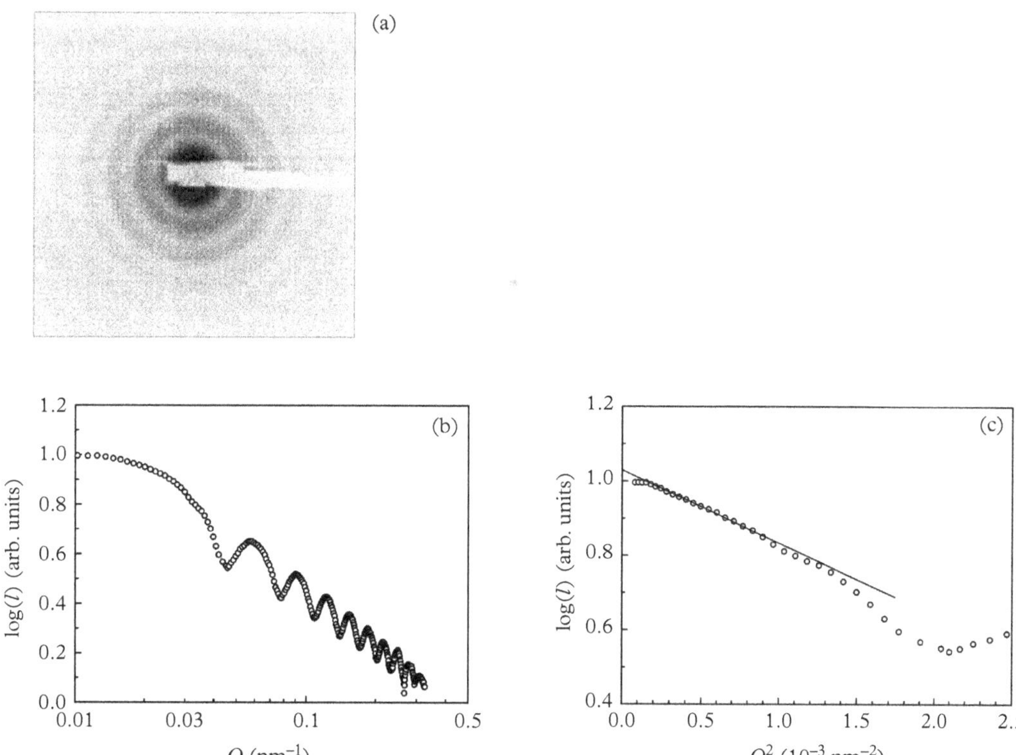

Figure 2.14 *SAXS data for a dilute suspension (0.1 volume %) of polystyrene spheres. (a) Two-dimensional detector pattern. (b) Double logarithmic plot after circular integration. (c) Guinier plot $\ln I(Q)$ vs. Q^2 (see Equation 2.14) of the low-Q region. (Adapted with permission from Megens et al., 1997. Copyright American Chemical Society.)*

We shall illustrate the behaviour discussed so far with SAXS experiments of a simple model system: a dilute suspension of (non-interacting) polystyrene spheres. Figure 2.14a shows the pattern observed at a two-dimensional detector, in which the concentric rings are due to the form factor. After circular integration, Figure 2.14b is obtained that can be directly compared with the model calculation of Figure 2.13. The first minimum is found at $Q = 0.0458\,\mathrm{nm}^{-1}$, leading to a radius $a = 98\,\mathrm{nm}$ (see the formula in Figure 2.13). This is in excellent agreement with the specification $a = 99\,\mathrm{nm}$ for the (commercial) polystyrene particles. A Porot plot gives $a = 101\,\mathrm{nm}$, but the slope deviates somewhat from the predictions. This is due to a slight dispersity[15] of 2.3% that could be derived from the observed deviations of ideal behaviour (Megens et al., 1997). Finally, in Figure 2.14c the low-Q data from Figure 2.14b are plotted against Q^2. The slope of the straight line has been calculated from Equation (2.14) using $-R_G^2/3 = -a^2/5$ with $a = 99\,\mathrm{nm}$, and is in excellent agreement with the data at low Q-values.

[15] Note the official term 'dispersity' instead of the probably better known 'polydispersity'. To be contrasted with 'uniform' systems.

2.4 Case study: Modelling SAXS of non-interacting particles

So far we have restricted ourselves to simple homogeneous systems with spherical symmetry and discussed how a model for the particles can be fitted directly to the scattering data. Obviously, shape, size, and structure of the particles will be reflected in the scattering curve. In this case study we shall first extend the method to more general shapes (form factors) like a thin disc and a long rod. For dilute systems (non-interacting particles), any shape anisotropy will be orientationally averaged and the scattering will still be spherically symmetric. We restrict ourselves to uniform systems and will not consider the complications due to dispersity. In general one can say that simple modelling will provide shape information if the system is known by other means to be uniform. Conversely, the dispersity can be determined if the shape is known, as discussed in the example at the end of Section 2.3. Interestingly, nature is capable of creating effectively uniform systems. In such cases, it is possible to fit the scattering to a model obtained by addition of form factors of simpler components (see later in this section).

Form factors for a large number of objects have been summarized by Pedersen (1997). The calculations given in Section 2.3 for spheres can be straightforwardly extended to a thin disc and a long rod. The results for the scattered intensity are shown in Figure 2.15. Most importantly, in the low-Q region of the form factor (Guinier regime), the exponent changes from the value of 0 for a sphere (see Figure 2.13) to -2 and -1 for a randomly oriented thin disc and long rod, respectively. These considerations can be extended to hollow and inhomogeneous particles (see Glatter, 2002). In Figure 2.16 we reproduce results from SAXS

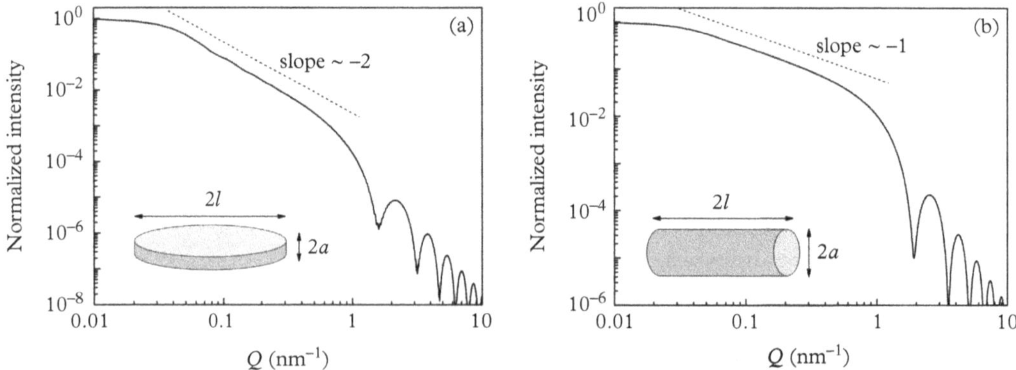

Figure 2.15 *Calculated SAXS pattern of non-interacting uniform colloid discs (a) and cylinders (b), respectively, for 2l = 100 nm and a thickness 2a = 4 nm. In the oscillatory Porod regime the exponent of the slope is still −4.*

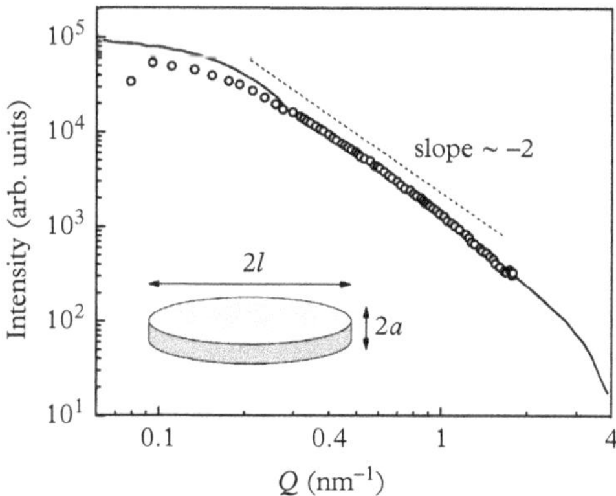

Figure 2.16 *Experimental SAXS results for a dilute suspension of discs (after Kroon, Vos, and Wegdam, 1998). The solid curve is calculated for uniform discs with diameter $2l = 25$ nm and thickness $2a = 2$ nm.*

experiments of a suspension of randomly oriented thin discs. At intermediate Q-values the intensity decays with Q^{-2} in agreement with the model calculations of Figure 2.15a.

During the last decade, great progress has been made in the analysis of SAXS from solutions of biological macromolecules. Unfortunately this topic is beyond the scope of the present volume; for a review we refer to Svergun (2007). Here, we want to extend the modelling of form factors discussed so far to the particular case of a phospholipid nanodisc. Such nanodiscs consist of a phospholipid bilayer disc, surrounded and stabilized by two amphiphilic protein 'belts'. The amino acid sequence of the proteins defines the length of the belt, which, in turn, controls the outer diameter of the disc. The confinement of a phospholipid bilayer in pieces of nanometer size makes them interesting as molecular sample holders for membrane proteins. Solutions of nanodiscs of DLPC[16] have been investigated in detail by SAXS and SANS. In the SANS experiments, the contrasts were varied by choosing two different concentrations of D_2O (Skar–Gislinge et al., 2010, 2011). The different contrasts in three types of measurement allowed different parts of the complicated structure to be highlighted. The SAXS results are shown in Figure 2.17. A structural model has been designed that fits to all

[16] DLPC stands for di-lauroyl-phosphatidyl choline.

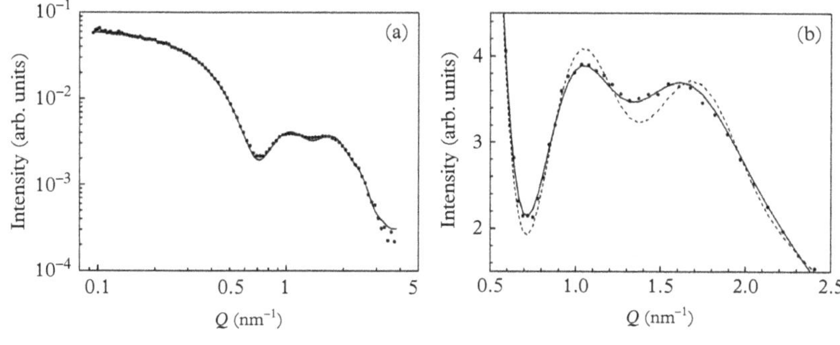

Figure 2.17 *SAXS pattern of DLPC. (a) Overall behaviour of fitting to a circular disc. (b) Details of fitting to a circular disc (broken line) and to an ellipsoidal disc (full line).*

Figure 2.18 *Calculating a complex form factor by subtraction of simpler ones.*

three measurements. It has been built by adding contributions from the different elements, while chemical and biophysical information is incorporated to constrain the number of possibilities (Skar–Gislinge and Arleth, 2011). We shall follow their line of thought and start with the basic elements.

Figure 2.18 illustrates how form factors of simple elements can be combined to more complex ones. In particular, the figure indicates how a 'belt' can be obtained by subtracting two cylinders of the same height but different diameter. In building the full model of Figure 2.19a, a whole series of form factor functions have to be combined (Figure 2.19b). The total scattering amplitude is written as the sum of the form factors of each disc weighted by the scattering length density of the appropriate group (see Box 2.1). We shall give a qualitative discussion of each of the terms in Figure 2.19b.

Phospholipid bilayer. The phospholipid interior of the nanodisc is represented by a stack of discs in which the top and bottom ones represent the hydrophilic head groups. To this, another pair of discs is added corresponding to the top and bottom alkyl layers. The system is completed by a middle layer representing the methyl end groups of the lipid tails, which are known to have a slightly lower electron density. Hence the total scattering of the bilayer is represented by the sum of the first three terms in Figure 2.19b.

Protein belt and his-tag. The protein belt stabilizing the lipid bilayer is taken as a hollow cylinder, like the one given in Figure 2.18. Inclusion leads to the top line of Figure 2.19b, and we arrive at a sum of four form factors

$$\Phi_{\text{nanodisc}} = \Phi_{\text{heads}} + \Phi_{\text{alkyl}} + \Phi_{\text{methyl}} + \Phi_{\text{belt}}.$$

It turns out that the model so far does not fit the experimental data in a satisfactory way. In fact, the protein belt is known to have a polyhistidine tag. This

Figure 2.19 *A model of DLPC (a) can be composed from a combination of form factors (b). (Adapted with permission from Skar–Gislinge and Arleth, 2010. Copyright American Chemical Society.)*

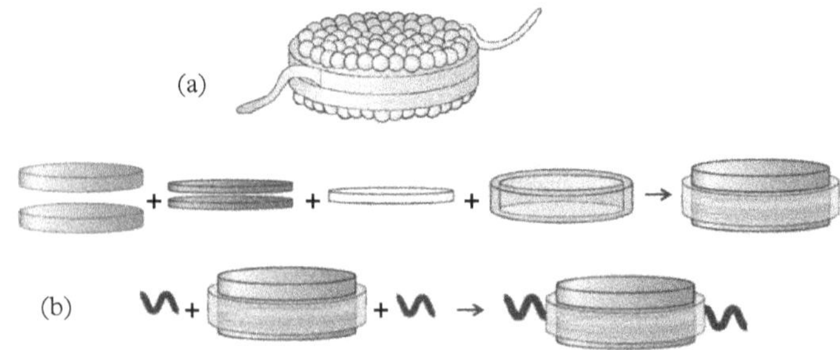

so-called his-tag consists of 22 amino acids and can be modelled as a Gaussian random coil. The his-tag is assumed to be uniformly distributed on the outer rim of the belts. Inclusion is crucial for a good fit of the data and leads to the final model of Figure 2.19b.

Interfacial roughness. The infinitely sharp interfaces in the model give rise to some nonphysical artefacts at the high-Q part of the scattering curves. To remove these, an average Gaussian distributed interface roughness has been introduced into the model.

Elliptical shape of the nanodiscs. For a disc fully loaded with phospholipids, one would a priori expect the nanodiscs to be of circular shape. One of the interesting results of the model is illustrated in Figure 2.17b. To fit the data one must assume that the nanodiscs are not circular but have a significant elliptical shape with an axis ratio of about 1.3. In agreement with statistical mechanical arguments and other evidence, the disc is not maximally loaded.

The detailed structural model of nanodiscs as described is a nice example of what is possible these days in direct modelling of scattering data. Nevertheless, in principle the approach suffers from an intrinsic difficulty of model fitting that we shall also encounter later in other chapters of this book: A model might provide an excellent fit to the data, but this does not necessarily prove its uniqueness. In the present case, much care has been taken to restrict the freedom of modelling by incorporating molecular constraints from biophysical and chemical information. In combination with the three different scattering contrasts, this results in a self-consistent model that is most probably very reliable.

2.5 Extension to simple liquids—Structure factor

Liquids consist of an assembly of molecules in space. To calculate their scattering we need the Fourier transform of this assembly, given by the sum of the Fourier transforms of the individual molecules. Evidently, we are not in the situation of a dilute solution anymore. Hence, taking the phase differences into account, interference between x-rays scattered by electrons in *different* particles also comes into play. We use the term particles instead of molecules because the same reasoning applies to any assembly of atoms ('super molecules' like, for example, a concentrated solution of colloids). Let us reconsider Equation (2.2), relating F^{mol} to a summation over the individual F^{atom}, which can be generalized for the present situation to

$$F(\mathbf{Q}) = \sum_{\mathbf{r}_j}^{\text{particles}} F_j^{\text{particle}}(\mathbf{Q}) \exp(i\mathbf{Q} \cdot \mathbf{r}_j).$$

In liquids and dense solutions (as in amorphous solids) any orientation of the particles will still be randomly distributed. Using the average value $\langle \exp(\mathbf{Q} \cdot \mathbf{r}_j) \rangle$ as calculated in Box 2.3, we can write

$$\langle F(\mathbf{Q}) \rangle = \sum_{\mathbf{r}_j}^{\text{particles}} F_j^{\text{particle}}(\mathbf{Q}) \frac{\sin(Qr_j)}{Qr_j}. \tag{2.16}$$

For spherically symmetric particles the scattered intensity is given by[17] $I(Q) = \langle F^2(Q) \rangle$, or

$$I(Q) = \sum_j \sum_k F_j(Q) F_k(Q) \frac{\sin(Qr_{jk})}{Qr_{jk}}$$

$$= \sum_j F_j^2(Q) + \sum_{j \neq k} \sum_k F_j(Q) F_k(Q) \frac{\sin(Qr_{jk})}{Qr_{jk}}.$$

If the system consists of N identical particles, we can rewrite this expression as

$$I(Q) = NF^2(Q) + F^2(Q) \sum_{j \neq k} \sum_k \frac{\sin(Qr_{jk})}{Qr_{jk}}$$

$$= NF^2(Q)S(Q), \tag{2.17}$$

with

$$S(Q) = 1 + \frac{1}{N} \sum_{j \neq k} \sum_k \frac{\sin(Qr_{jk})}{Qr_{jk}}.$$

Equation (2.17) is similar to the first part of Equation (2.11), apart from the presence of the factor $S(Q)$. The latter takes the *inter*particle interactions into account and is called *structure factor* or *interference function*. In the simplest case we suppose that each particle in the sample has the same average relationship with its neighbours. Then each of the N particles interacts with $(N-1)$ neighbours, leading to $N(N-1)$ identical interactions. For the large number of particles involved $N \approx N-1$ and we can write

$$S(Q) = 1 + N \sum_k \frac{\sin(Qr_k)}{Qr_k}. \tag{2.18}$$

Note that in this equation r_k is the distance of the kth particle from the centre of some arbitrarily chosen one, which has become the centre of the coordinate system. The meaning of the structure factor is once more illustrated in Figure 2.20. Obviously, $S(Q)$ will become more complicated if there is correlation between positions and orientations.

In fact, life is not that simple, and the assumption of the same relationship of a molecule with its neighbours has to be refined. For that purpose, the radial distribution function $g(r)$ can be introduced, which gives the probability to find a

[17] This is a special case of the general expression $I(Q) = F(Q)F^*(Q)$ in which $F^*(Q)$ is the complex conjugate.

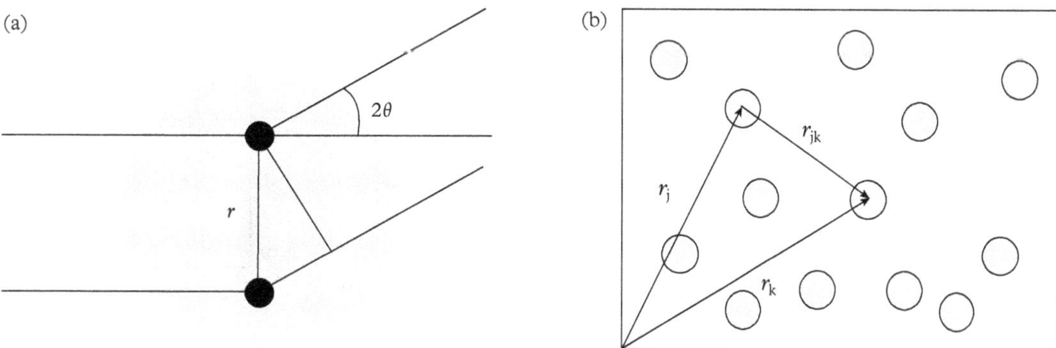

Figure 2.20 *Illustration for a non-diluted system of spheres of (a) the form facture $\Phi(Q)$ describing the interparticle size and shape, and (b) the structure factor $S(Q)$ which is a function of the local order and the interaction potential.*

molecule at some distance r from the origin.[18] The appropriate volume element is $4\pi r^2\,dr$ and, denoting the total volume by V, the normalized probability of finding a molecule in this volume is

$$(4\pi r^2\,dr/V)g(r).$$

This factor should be incorporated into Equation (2.18). As $g(r)$ is a continuous function, the summation can be replaced by an integral. Defining $n_0 = N/V$, Equation (2.18) can be written as

$$S(Q) = 1 + n_0 \int_0^\infty 4\pi r^2 g(r)\frac{\sin(Qr)}{Qr}\,dr.$$

Replacing $g(r)$ by $g(r) - 1 + 1$, we arrive at

$$S(Q) - 1 = n_0 \int_0^\infty 4\pi r^2 [g(r) - 1]\,\frac{\sin(Qr)}{Qr}\,dr + n_0 \int_0^\infty 4\pi r^2 \frac{\sin(Qr)}{Qr}\,dr$$

$$= n_0 \int_0^\infty 4\pi r^2 [g(r) - 1]\,\frac{\sin(Qr)}{Qr}\,dr.$$

In the last equality we use $\sin(Qr)/(Qr) \to 0$ for $r \to \infty$. By writing $S(Q)$ in this form we can easily see that $[g(r) - 1]$ is related to $[S(Q) - 1]$ by inverse Fourier transformation,

$$g(r) - 1 = (1/n_0) \int 4\pi r^2 [S(Q) - 1]\,\frac{\sin(Qr)}{Qr}\,dQ. \tag{2.19}$$

This allows x-ray scattering to determine the radial distribution function from which, in turn, theories of atomic or molecular liquids can be tested.

[18] Note that we still retain spherical symmetry by choosing $g(r)$ to depend only on the radial coordinate r.

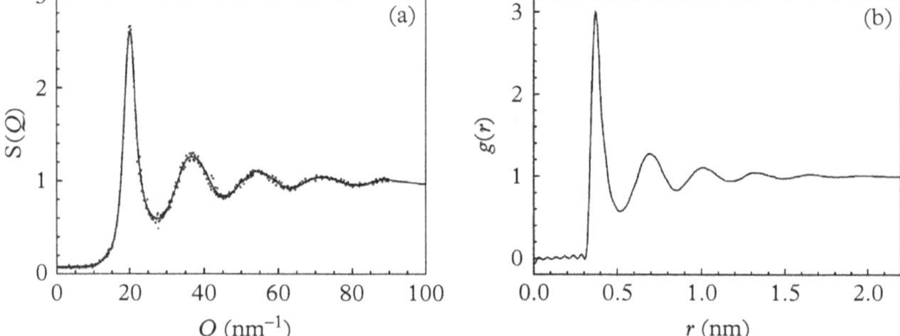

Figure 2.21 *Interference function (a) and resulting radial distribution function (b) of liquid argon at 85 K (after Yarnell et al., 1973).*

In Figure 2.21 we show experimental results from neutron scattering for liquid argon. From measurements of $I(Q)$ according to Equation (2.17), $S(Q)$ is obtained as displayed. The full line represents a smoothed and extended representation of the data. This curve forms the basis for the calculation of $g(r)$ according to Equation (2.19). The result is in excellent agreement with molecular dynamics calculations using a Lennard–Jones potential for argon. The latter coincides with Monte Carlo calculations using a more sophisticated potential describing the intermolecular interactions. However, the results deviate from computations using simply hard-sphere interactions.

Finally, Figure 2.22 summarizes the various cases considered so far: essentially, isotropically distributed uniform spheres at different densities. The results also apply qualitatively to less precisely defined systems that can be approximated by spheres. However, it should be realized that the validity of the concepts of $S(Q)$ and $g(r)$ are not restricted to spherical symmetry. Quite generally we can define $S(Q)$ as $\langle \exp(-i\mathbf{Q} \cdot (\mathbf{r}_j - \mathbf{r}_k)) \rangle$, which reduces to $\sin(Qr_k)/(Qr_k)$ in the spherical case. Similarly, the present radial distribution function $g(r)$ is the spherical equivalent of the general pair distribution function $g_2(\mathbf{r}_j, \mathbf{r}_k)$ (see Roe, 2000).

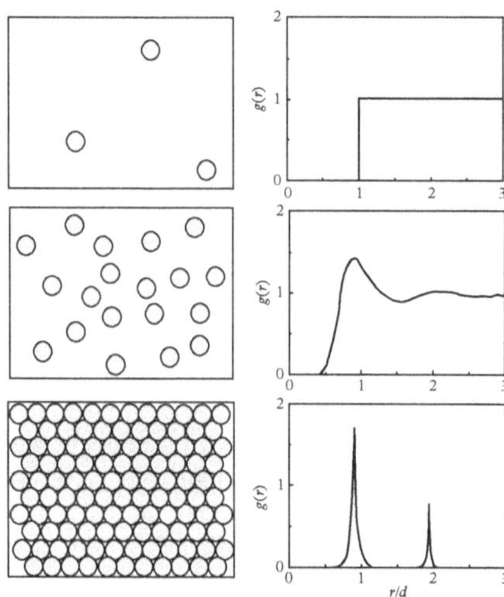

Figure 2.22 *Radial distribution function $g(r)$ for isotropically distributed uniform spheres at different densities.*

Order/Disorder in Soft Matter

3

This chapter is, to some extent, an intermezzo. Having discussed essentially disordered systems in the previous chapter, and before coming to ordered crystals in the next one, we treat the various types of order that are encountered in soft matter. As we shall see, a rich spectrum of possibilities exists, comprising short-range order, long-range order, and intermediate situations. In addition to positional order—implicitly meant so far—in the case of anisotropic molecules orientational order also comes into play. These possibilities are especially important for present-day developments in organic chemistry like self-assembling systems. We shall use, extensively, liquid crystals as model systems, because their anisotropy serves well to demonstrate many of the points mentioned.

3.1 Short-range order

At the end of Chapter 2 we discussed the radial distribution function of liquid argon: the probability of finding an argon atom at a specific position surrounding a central one (Figure 2.21b). There is an excluded region close to the central atom, then an increased density for close neighbours, followed by an oscillating decrease in correlations towards the average density at a couple of molecular dimensions. This type of short-range correlations is quite general. For non-spherical molecules, obviously, directional effects will also play a role. In addition, apart from the excluded volume effect (repulsion) considered for argon, various types of force like Van der Waals forces, ionic forces, or hydrogen bonding may occur.

A special anisotropic example is water. The oxygen in a water molecule bonds its two hydrogen atoms at an angle of 105° and arranges the four other electrons in two lone pairs. The liquid gains attractive energy by pointing the negative lone pairs towards positive hydrogen atoms (hydrogen bonding). The water molecules try to form chains or clumps in which oxygen atoms are tetrahedrally arranged but in which the twisted dumbbell molecules at the same time do not overlap. Though overall, again, a spherical average is obtained, this is not true locally. In fact, three radial distribution functions must be considered—displayed

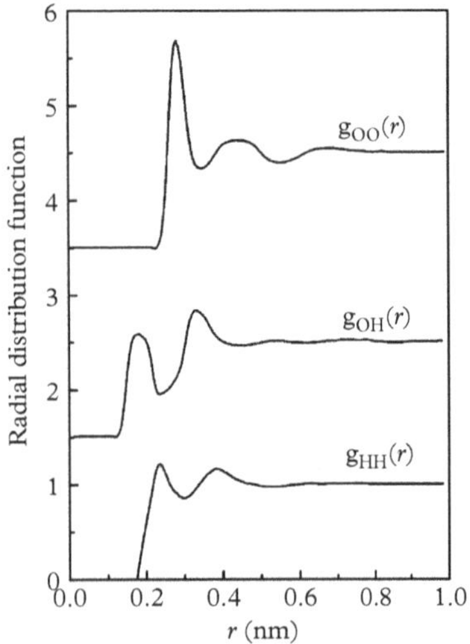

Figure 3.1 Radial distribution functions of liquid water at room temperature (Soper, Bruni, and Ricci, 1997). Curves are shifted for clarity.

in Figure 3.1—corresponding to the local OO, OH, and HH interactions. Obviously, $g_{OO}(r)$ is the most important one with a peak at 0.28 nm corresponding to the well-known average hydrogen bond length.

Short-range order plays a role in all liquids, gasses, and amorphous solids. These systems are often, on average, isotropic: they have a random distribution of the centres of mass that can be well described by mean-field theory. The averaging process can be either static or dynamic. However, to understand the local (short-range) structural behaviour, more detailed models are required. For simple liquids obviously (hard-sphere) repulsions are important in combination with attractive Van der Waals forces. A common model incorporating both effects is the so-called Lennard–Jones potential.

The local density oscillations can be described by a periodic function with exponentially decaying amplitude. The extent of short-range order is determined by a correlation length ξ corresponding to the 1/e value of that amplitude. The Fourier transform of an exponential function is a Lorentz function (see Figure 3.2), which is characterized by relatively long tails. Such a Lorentzian can often be well fitted to x-ray results for the scattering of a liquid-like structure. In the present examples, ξ had relatively small values (a couple of molecular diameters). In modern self-assembling chemistry sometimes cases occur of short-range order over a more extended range. We still classify such orders as 'short-range' if the scattering can be described in terms of a Lorentz function with a typical correlation length, ξ. This is the defining difference from long-range orders.

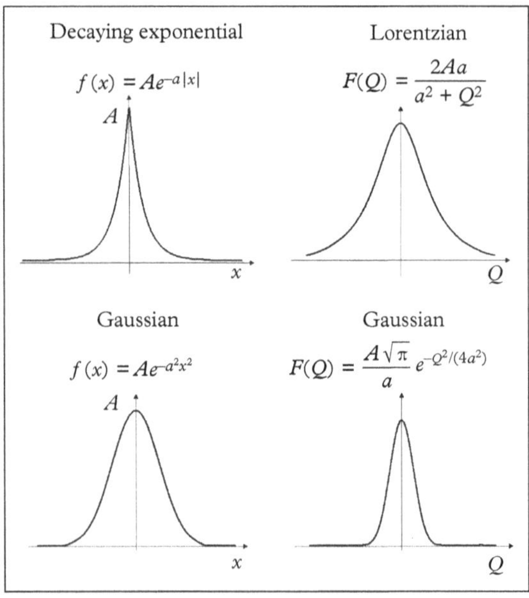

Figure 3.2 Exponential function (symmetrized) and Gaussian function, and their Fourier transform.

3.2 Long-range order

Long-range order is what we usually have in mind when discussing x-ray scattering in terms of relatively narrow peaks found at certain positions in a crystal. Qualitatively, the description is simple; let us recall Equation (2.17),

$$I(Q) = NF^2(Q)S(Q).$$

The structure factor (interference function) $S(Q)$ is essentially determined by the probability of finding a scatterer at a particular point in space. In the case of a crystal, it will be strongly peaked in certain directions that coincide with lattice planes.[1] This leads to sharp x-ray peaks in the total scattered intensity. These can usually be fitted to a Gaussian function; to be contrasted with the much broader tails of a Lorentzian. Note that the scattering from a Gaussian function is again a Gaussian, though with a different width (see Figure 3.2). Figure 3.3 shows two examples from smectic liquid-crystalline elastomers,[2] for which the nature of the organization of the layer structure could be derived from the shape of the x-ray peaks.

Long-range order is theoretically unlimited. Each subsequent lattice point in space (up to arbitrarily large distances) can be determined from the previous sequence. In other words: the relevant correlation function is essentially constant. However, in practice, long-range order is not necessarily very long. In any real sample there will be crystallites of a certain finite size. This is even true for a perfect single crystal, for which the sample size is the crystallite size. As a result, the structure factor $S(q)$ will display finite-size broadening and not be sharply peaked anymore. For nano-crystals even sidebands can occur. This effect is illustrated in Figure 3.4. Physically it is similar to the situation observed for a finite-size optical grating that shows narrowing of the central peak when the number of scratches is increased.

[1] Discussed in more detail in Chapter 4.

[2] The nature of smectic phases and smectic elastomers is discussed in Section 3.3.2. For the moment we are only concerned with the difference in line shape that reflects a qualitative difference in order.

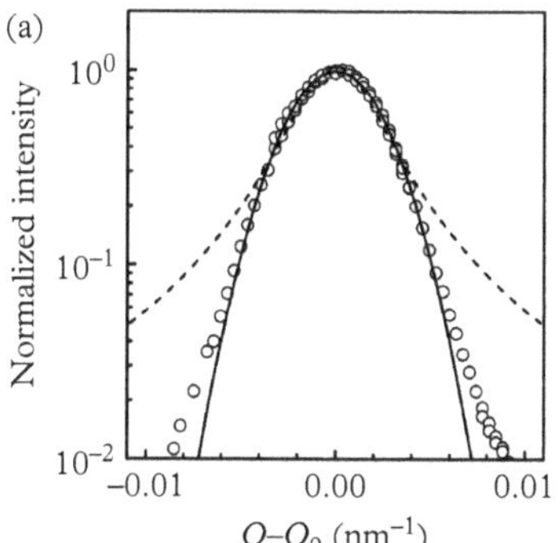

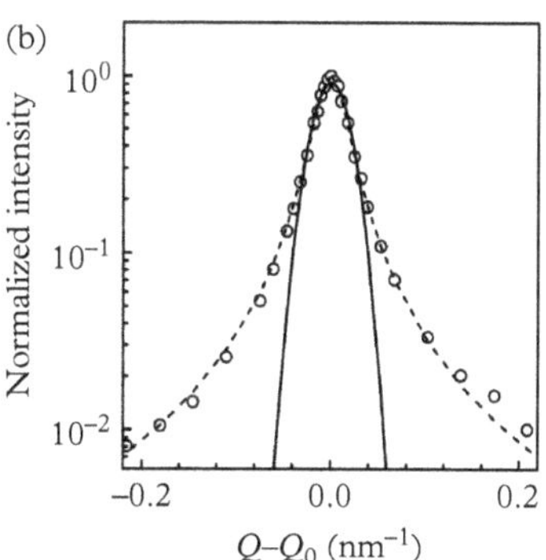

Figure 3.3 *Normalized x-ray peaks of two smectic elastomers with (a) weak and (b) strong crosslinking.*
Note the logarithmic vertical scale, indicating two orders of magnitude in the measured intensity. Full lines are best fits to a Gaussian function and broken lines to a Lorenzian one. The difference in horizontal scale reflects the different width of the Gaussian peak in (a) and the broad Lorentzian peak in (b) (Obraztsov et al., 2008).

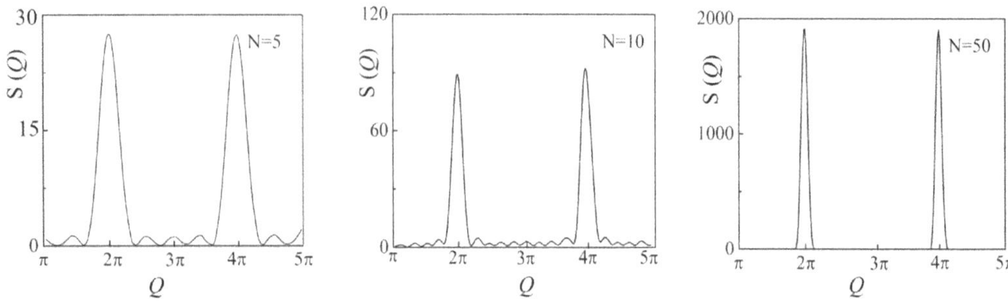

Figure 3.4 *Structure factor $S(Q)$ for a different number of periodicities N.*

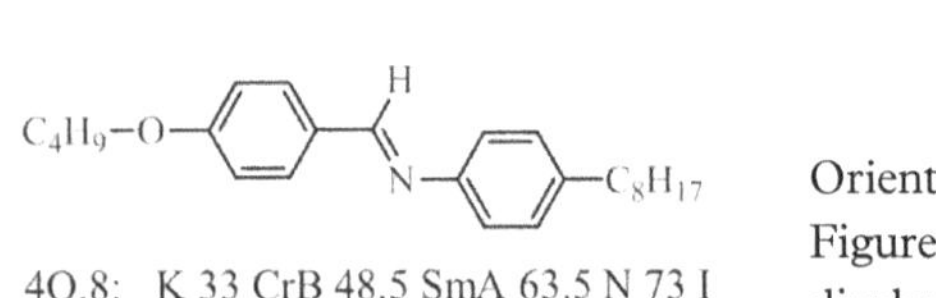

4O.8: K 33 CrB 48.5 SmA 63.5 N 73 I

7AB: K 32.4 SmA 53.5 N 70 I

Figure 3.5 *Molecular structure and phase behavior of the compounds indicated as 4O.8 and 7AB, respectively. Transition temperatures are in °C.*

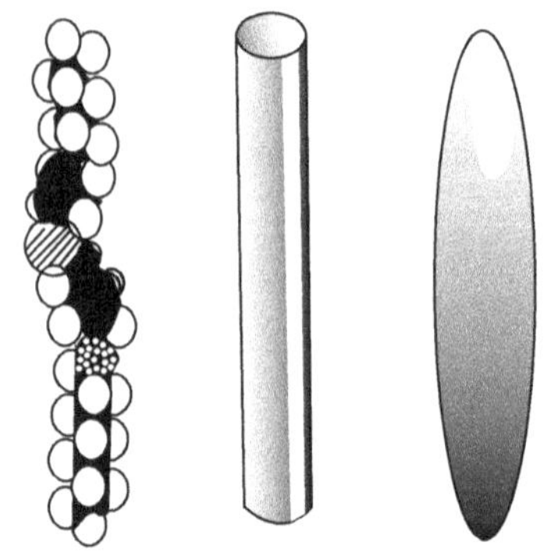

Figure 3.6 *Representations of a mesogenic molecule.*

3.3 Orientational and positional order in liquid crystals

Orientational order is particularly important in the context of liquid crystals. Figure 3.5 shows some typical mesomorphic (liquid-crystalline) molecules that display, upon cooling from the high-temperature isotropic liquid (I), first a nematic phase (N), then a smectic one (SmA), and finally one or more crystalline phases (CrB and K). These are characteristic classical molecules with liquid-crystal phases. Nowadays many more exotic molecules have been shown to display liquid-crystalline order. Typically, the shape of such a molecule can be represented by a rod or ellipsoid (see Figure 3.6). Note that any rotational symmetry around the long molecular axis can only be due to dynamic rotational averaging. It will be clear that in the isotropic phase the short-range order must be anisotropic. Two neighbouring molecules will have an enhanced probability to be parallel, which tendency will again decay exponentially. Hence there must be a strong coupling between short-range orientational and short-range positional order, the description of which is not simple (contrary to the situation for liquid argon). Note that short-range orientational and positional order is not restricted to liquid crystals, but can be found in all types of system, for example, also in polymers, which is not always fully appreciated.

3.3.1 The nematic phase

The nematic phase is characterized by long-range orientational order, while the centres of mass are—on average—still distributed at random (see Figure 3.7, where the orientational order parameter S is also defined). $S = 1$ for perfect alignment of the long molecular axes along the preferred direction **n** (the director) and $S = 0$ in the isotropic state. Note that there is up-down symmetry: **n** and −**n** are equivalent as expressed in the quadratic function for the order parameter (a linear function would give $\langle \cos \beta \rangle = 0$). Figure 3.8a shows an x-ray picture of a nematic phase in which the director is arbitrarily oriented inside the sample. We can

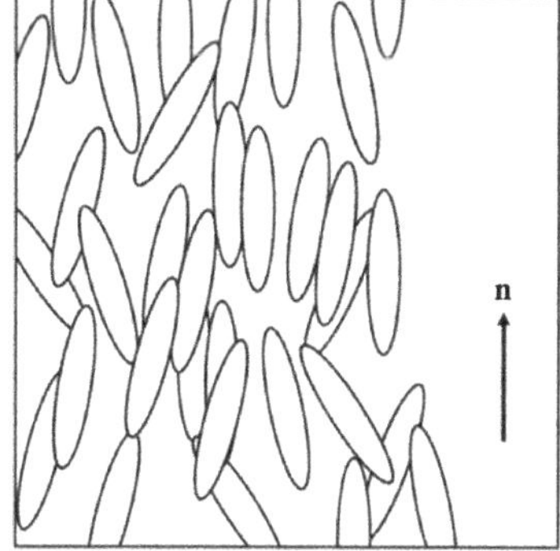

Nematic phase:
Orientationally ordered molecules
Random distribution centres of mass

Orientational order parameter:
$S = \frac{1}{2} \langle 3\cos^2 \beta - 1 \rangle$

Figure 3.7 *Nematic liquid-crystal phase with order parameter S.*
Here, β is the angle between the long axis of a molecule and the director **n** *along the z-axis while the brackets indicate an average.*[3]

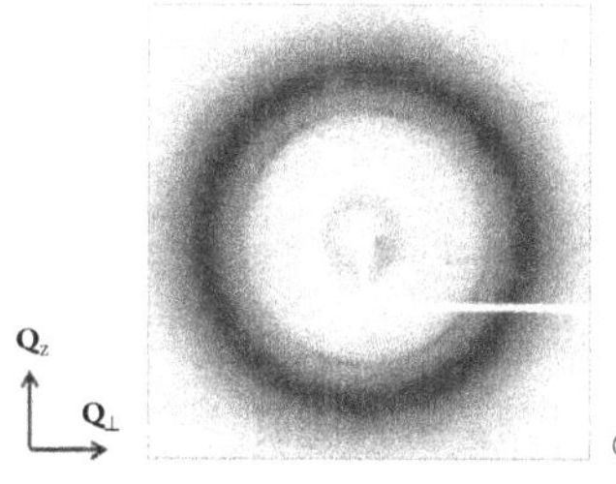

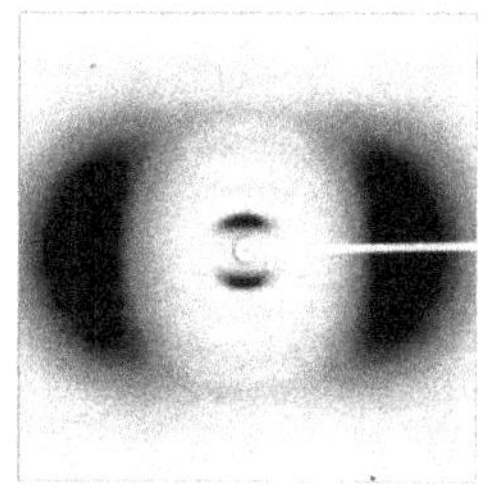

Figure 3.8 *X-ray pattern of a nematic phase. (a) Randomly oriented. (b) Uniformly oriented with* **n** *along* Q_z *(after Leadbetter, 1979).*

distinguish two rings. The large one corresponds to short-range order perpendicular to the director. Hence, on average, it is perpendicular to the long axis of the molecules and corresponds to the small dimension. In addition, another less intense ring is visible at small Q-values, just outside the central beam stop. This corresponds to a substantially larger period, matching the long dimension of the molecules. Both rings are broad and diffuse, as expected for short-range order.

The director can be made uniform (for example by an external magnetic field), which allows us to directly measure, by x-ray scattering, the anisotropy of the short-range positional order (Figure 3.8b). These data are remarkable in showing aligned short-range order in two different directions as evidenced by broad x-ray peaks, parallel and perpendicular to **n**.

3.3.2 The smectic-A phase

The smectic-A phase consists of 'stacks' of liquid layers (Figure 3.9).[4] The molecules in the smectic planes form a two-dimensional liquid with only short-range positional order. Perpendicular to the layers there is a (weak) density modulation, and the description as 'stacks' or 'layers' is convenient but far too strong. Nevertheless, this density wave can be considered as the simplest form of melting/crystallization in one direction only.

[3] An alternative notation is $S = \langle P_2(\cos\beta) \rangle$ in which P_2 is the second-order Legendre polynomial.

[4] We shall encounter smectic phases with ordered layers in Section 3.4.2.

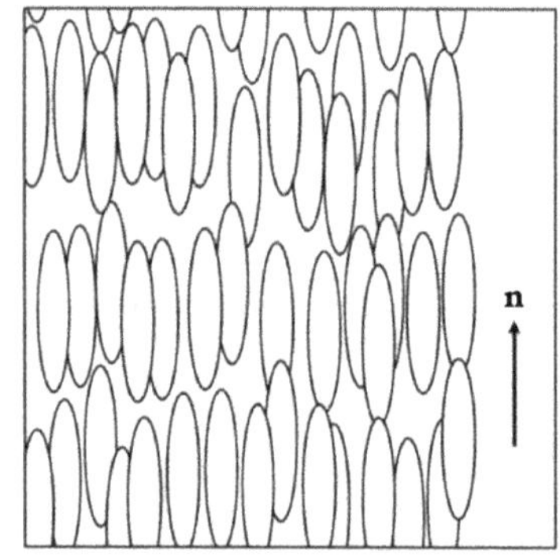

Smectic phase:
Orientationally ordered molecules
Stacked liquid layers with period d

Figure 3.9 *Smectic-A   liquid-crystal phase.*
The z-axis is the direction along **n.**
For simplicity, the orientational order is assumed to be perfect (S = 1).

Smectic order parameter:
$\tau = \langle \cos Q_0 z \rangle$ with $Q_0 = 2\pi/d$

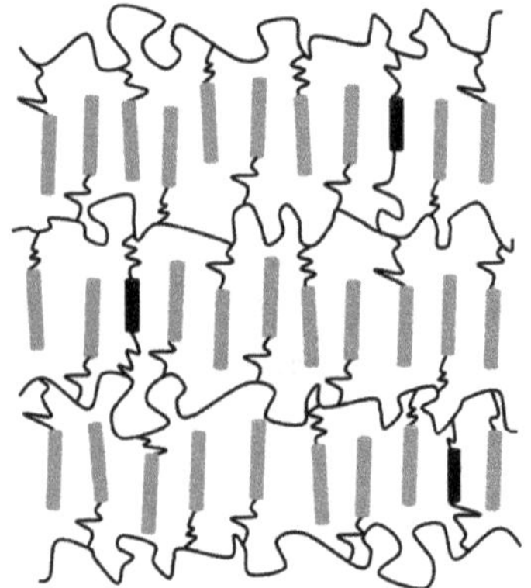

Figure 3.10 *Schematic picture of a smectic side-chain elastomer.*
The side groups are indicated in dark grey, the crosslinks in black.

Interesting variations on the simple smectic phase of Figure 3.9 are smectic polymers and elastomers. In the simplest case the smectogenic groups are attached to a polymer chain via flexible spacers (see Figure 3.10). The latter groups are important to keep the flexibility needed to form still smectic layers. In a next step, the polymer chains can be connected by crosslinks (with a diverse degree of flexibility), leading to smectic elastomers. We do not go into further detail, but note that in Figures 3.3 and 3.14 smectic elastomers are used as an example to discuss some x-ray properties of smectic phases.

X-ray patterns of a smectic-A phase (liquid layers) are shown in Figure 3.11. Figure 3.11a shows two rings, at large and at small angles, respectively. The picture of an oriented sample (Figure 3.11b) is more informative. We see along $Q_\perp$ still a broad liquid peak as in the nematic phase. However, along the layer normal (direction of Q_z), we note sharp peaks corresponding to the layer structure. Also in this particular case, a second-order of the main layer peak is visible. After circular averaging, the diffractogram of Figure 3.11c is the result. A similar

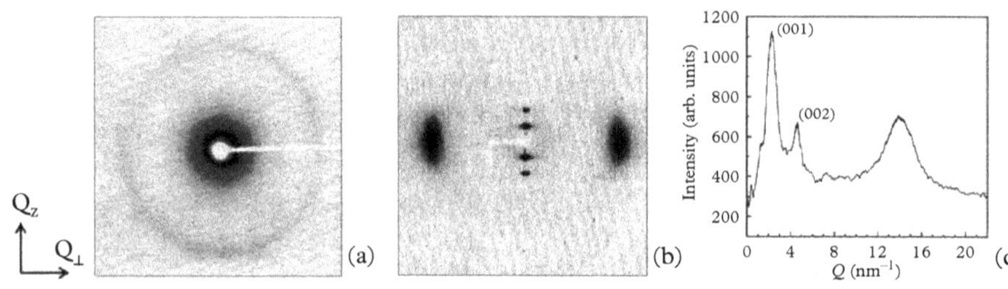

Figure 3.11 *X-ray pattern of a smectic-A phase. (a) Randomly oriented. (b) Uniformly oriented. (c) Circular averaged.*
The fundamental smectic layer peak (001) at $Q = 2.28\,nm^{-1}$ corresponds to $2\pi/Q = 2.8\,nm$; the liquid in-plane order at $Q = 14\,nm^{-1}$ to 0.45 nm (approximately the length and width of the molecules, respectively).

picture would be obtained after averaging the randomly oriented sample of Figure 3.11a. The interpretation of the smectic one-dimensional density wave is of considerable interest. Are standard ideas about crystallization still applicable to this special situation? The answer is no. As we shall discuss, because of its low-dimensional nature, the smectic layers are strongly susceptible to thermal fluctuations, contrary to a three-dimensional crystal.

The properties of a smectic-A phase can be described using the Landau–de Gennes theory. Without going into details we shall outline the general aspects. The theory considers two types of deformation, $u(z)$, of the smectic layer positions along the layer normal, due to undulations and to compression/dilatation of the layers, respectively. See Figure 3.12 in which L is a finite size along the z-direction while large perpendicular dimensions are assumed. The first deformation is associated with an orientational bending elastic constant K and a wave vector $Q_\perp$, the second one with a 'classical' solid-like elastic constant B and a wave vector Q_z. Bending of the (liquid) layers costs little energy as their thickness is maintained: K has a small value, contrary to B. Taking these deformations into account, the Landau–de Gennes free energy can be given as

$$F = \frac{1}{2} \int d^3 r \left\{ B \left(\frac{\partial u(\mathbf{r})}{\partial z} \right)^2 + K \left[\frac{\partial^2 u(\mathbf{r})}{\partial x^2} + \frac{\partial^2 u(\mathbf{r})}{\partial y^2} \right]^2 \right\}. \tag{3.1}$$

The free energy can be rewritten as a sum of contributions from compressional and bending modes proportional to BQ_z^2 and $KQ_\perp^4$, respectively. Using the equipartition theorem, $\frac{1}{2}k_{\mathrm{B}}T$ can be assigned to each independent mode. The integration limits in the relevant z-direction are given by the layer period d and the sample size L. Then the result for the mean-square fluctuation amplitude can be calculated as

$$\langle u^2(z) \rangle = \frac{k_B T}{8\pi \sqrt{KB}} \ln \left(\frac{L}{d} \right). \tag{3.2}$$

As we see, the fluctuations diverge with increasing L—the size of the sample along the layer normal—though very slowly because of the logarithm. Hence, no

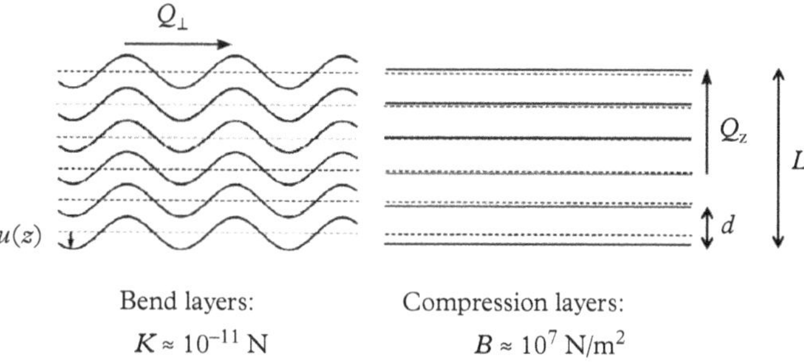

Figure 3.12 *The two types of deformation of smectic layers.*

pure long-range order is involved, which would essentially continue indefinitely. Instead the logarithmic divergence of fluctuations leads to so-called 'quasi-long-range order'[5] that decays algebraically as $r^{-\eta}$ (η small positive). The parameter η is given by the prefactor in Equation (3.2), determined by the two elastic constants, or more precisely

$$\eta = \frac{Q_0^2 k_B T}{8\pi \sqrt{KB}} \text{ with still } Q_0 = 2\pi/d. \tag{3.3}$$

The different types of order discussed so far in this chapter are summarized in Figure 3.13 together with their x-ray signature. Theoretically long-range order would give an infinitesimally sharp peak (delta-function). In practice it will be broadened by finite-size effects (as discussed in Section 3.2) and by the instrumental resolution, leading to a Gaussian-type line shape. As the tails of a Gaussian function vary as Q^{-2}, the difference with algebraic decay $Q^{-2+\eta}$ (Equation 3.5) are minor for a first-order peak with $n = 1$ and small values of η. The difference shows up more clearly for higher orders n for which algebraic decay leads to a line shape decaying with a power $(-2 + n^2\eta)$.

Figure 3.14 gives an analysis of the line shapes of x-ray peaks of a smectic elastomer of the type in Figure 3.10, used in Figure 3.3a.[6] As demonstrated there,

[5] The name refers to the small and subtle difference with 'true' long-range order.

[6] We restrict ourselves to Q_z, normal to the smectic layers. A more general discussion of the x-ray properties of smectic elastomers has been given by de Jeu and Ostrovskii (2012).

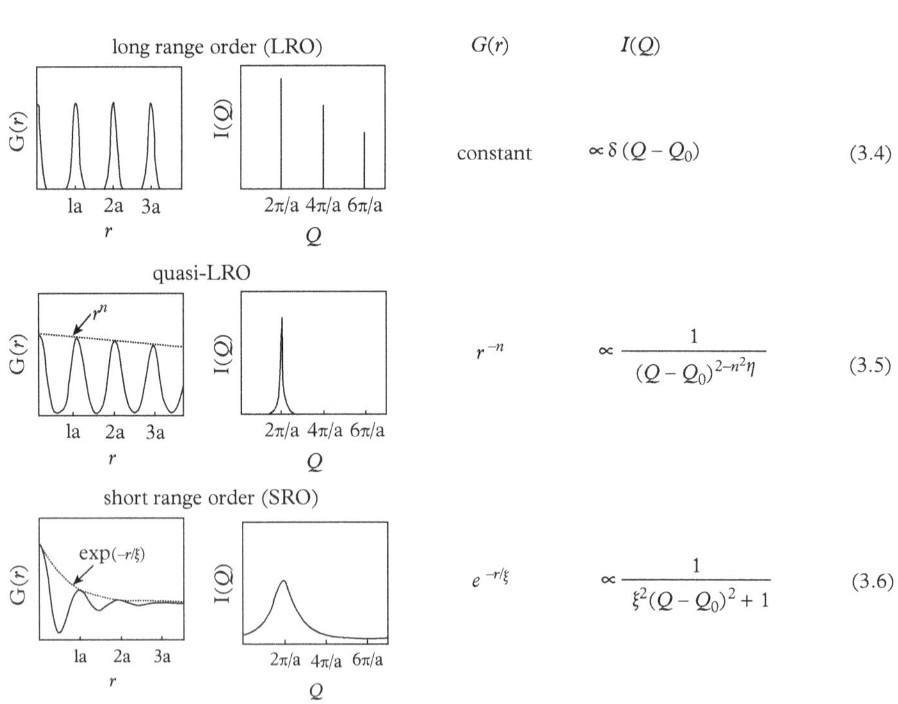

$$\propto \delta(Q - Q_0) \tag{3.4}$$

$$\propto \frac{1}{(Q - Q_0)^{2-n^2\eta}} \tag{3.5}$$

$$\propto \frac{1}{\xi^2(Q - Q_0)^2 + 1} \tag{3.6}$$

Figure 3.13 *Illustration of the various types of order and their mathematical description.*

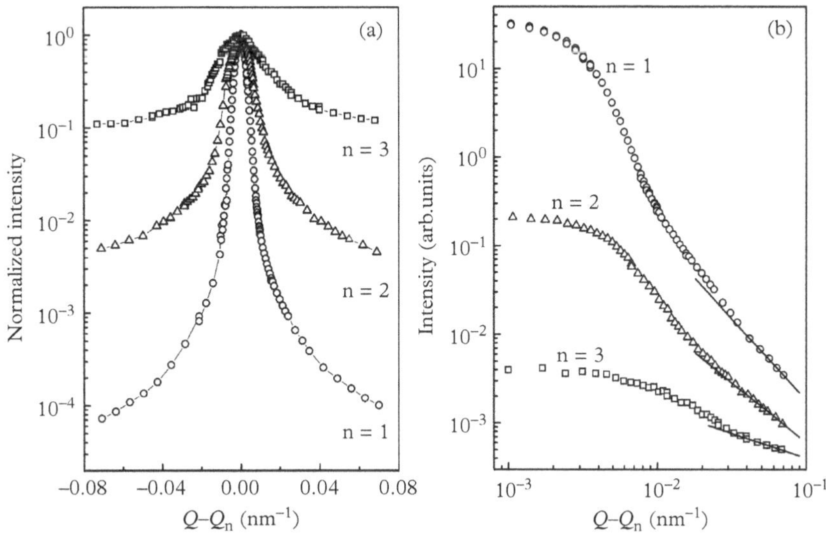

Figure 3.14 *(a) Three orders of line shape of a smectic elastomer. Note for n = 1 the four orders of magnitude in intensity compared to two orders in Figure 3.3, providing detailed information about the tails of the peak. (b) Logarithmic plot in Q indicating that the scaling relation $2 - n^2\eta$ for algebraically decaying quasi-long-range order holds for $\eta = 0.16$ (straight lines at large Q), (Obraztsov et al., 2008).*

the central part can be described by a Gaussian. However, the more complete measurement of Figure 3.14 indicates that the tails of the peak extend to large Q-values due to the algebraic decay. This is especially true for the higher-order peaks. According to Equation (3.5), for large Q-values the line shapes for the orders $n = 1, 2, 3$ decay as $2 - \eta$, $2 - 4\eta$, and $2 - 9\eta$, respectively. In the double logarithmic plot of Figure 3.14 this shows up nicely as straight lines at large Q that obey the scaling relation for an exponent $\eta = 0.16$. The fulfilment of this scaling proves satisfactorily that the smectic layer order indeed decays algebraically.

3.4 Order and dimensionality

3.4.1 Fluctuations at low dimensions

The peculiarities of smectic layer order are but one example of the general influence of the dimensionality of a system on its ordering. Let us consider, more generally, order in three-, two-, and one-dimensional space. Of course, we live in three dimensions, but two-dimensional effects come into play if we consider surface effects, monolayers, etc. The point is that potentially destabilizing thermal fluctuations are increasingly important in low-dimensional systems. Consider a crystal of typical size L, volume L^3, with a density function $\rho(x, y, z)$. This function is periodic with certain symmetry properties (see Chapter 4). Why is such a crystal stable, at least in three dimensions? The answer is that any fluctuations around lattice points remain finite.

Let us study the stability criterion in more detail in N-dimensional space ($N = 1, 2, 3$). The way to investigate this problem is to postulate a spontaneous (thermal) fluctuation and investigate whether it remains finite or diverges. Let $u_k(\mathbf{r})$ be such a fluctuation with amplitude u_k depending on the wave number $k = 2\pi/\lambda$. The wavelength spectrum allowed for such fluctuations is limited as follows

$$\text{minimum } \lambda : \text{ molecular dimension } a \to k_{\max} = 2\pi/a$$

$$\text{maximum } \lambda : \text{ typical sample size } L \to k_{\min} = 2\pi/L.$$

We use Hooke's law for the energy f_k associated with a fluctuation mode (harmonic approximation leading to a quadratic form of the free energy):

$$f_k = \frac{1}{2} C k^2 u_k^2$$

in which C is an elastic constant.

Using again the equipartition theorem, we attribute to each mode an average energy $\frac{1}{2} k_B T$. Making a summation over all modes in the N-dimensional volume $V = L^N$, one can calculate the average total fluctuation amplitude as

$$\sum_k \langle u_k^2 \rangle \propto \frac{k_B T}{VC} \int_{2\pi/L}^{2\pi/a} k^{-2} d^N k. \tag{3.7}$$

The importance of this result is not in the details of the formula, but in the behaviour of the integral as a function of N. Depending on the dimension $N = 1, 2, 3$, we integrate k^{-2} once, twice, or three times. Taking into account that a is a small quantity and that the sample size L can increase at liberty, we arrive at the results of Table 3.1. As we see, a three-dimensional lattice is stable against fluctuations because any fluctuation remains finite. In contrast, a one-dimensional lattice cannot exist and fluctuates away with increasing size L.

Table 3.1 *Integral in Equation (3.7).*

N	Integral	Limit for $L \to \infty$
1	$L/2\pi - a/2\pi$	$L/2\pi$
2	$2\pi \ln(L/a)$	$2\pi \ln(L/a)$
3	$8\pi^2/a + 8\pi^2/L$	$8\pi^2/a$

The most interesting situation is a two-dimensional lattice.[7] It cannot have long-range order as the lattice points are not defined for $L \to \infty$. However, for any realistic finite-size sample the fluctuation amplitude will still be small due to the slow logarithmic divergence. Though this general two-dimensional situation differs from the smectic layering discussed above (one-dimensional order in a three-dimensional system), the resulting logarithmic divergence of the fluctuations is similar.

A nice example of two-dimensional fluctuations is provided by capillary waves travelling on a liquid interface, say water. The roughness of the liquid interface leads to an average height profile to which two factors contribute. The first one is the intrinsic interface width, which is of the order of magnitude of the size a of the molecules. The second one is due to thermally excited capillary waves. We can describe each contribution by a Gaussian function, and add the width of the two terms quadratically. Using the theory of capillary waves[8] we can then write in a simplified way

$$\sigma_{\text{tot}}^2 = \sigma_0^2 + \sigma_{\text{cw}}^2$$
$$= \sigma_0^2 + \frac{k_\text{B} T}{2 \pi \gamma} \ln \left(\frac{Q_{\max}}{Q_{\min}} \right). \tag{3.8}$$

Here, σ_0 refers to the intrinsic width and γ is the surface tension. The limiting wavelengths are $Q_{\max} = 2\pi/a$, while $Q_{\min}$ is inversely proportional to the surface area that can be very large (lake, ocean). Evidently we meet again a logarithmic divergence of fluctuations, now as a consequence of the two-dimensional nature of the surface problem. As a result of Equation (3.8), on a large area like an ocean capillary waves could rise very high! In practice this does not happen: gravity was not included so far and sets a long-wavelength cut-off equal to $Q_{\min} = \rho g / \gamma$ in which ρ is the density and g the gravity constant.[9] X-ray reflectivity experiments of capillary waves will be discussed in Section 6.1.3.

3.4.2   Two-stage melting in two dimensions

There is still another aspect of smectic liquid crystals as two-dimensional systems. Let us go back to the model of smectic-A as a system of stacked liquid layers (Figure 3.9). Upon cooling we can expect the individual liquid layers to solidify. If the molecular positions between different layers are coupled, this would be nothing but a three-dimensional crystal. In fact, most of the liquid-crystal phases with stacked solid layers (in the older literature often indicated as smectic-B) are such lamellar crystals. Nowadays these systems are designated as crystalline-B (CrB as in Figure 3.5). However, if the coupling between the smectic layers is absent or at least weak, solidification of the liquid layers in smectic-A would correspond to two-dimensional crystallization (or reversely melting) in each of the layers.

[7] The limiting case is called the marginal dimensionality of the system, in this case 2.

[8] See, for a review, Penfold (2001).

[9] To be more precise: The restoring forces for thermally excited capillary waves are surface tension and gravity. The latter is dominant for long waves on the ocean, the former for ripples on a cup of tea.

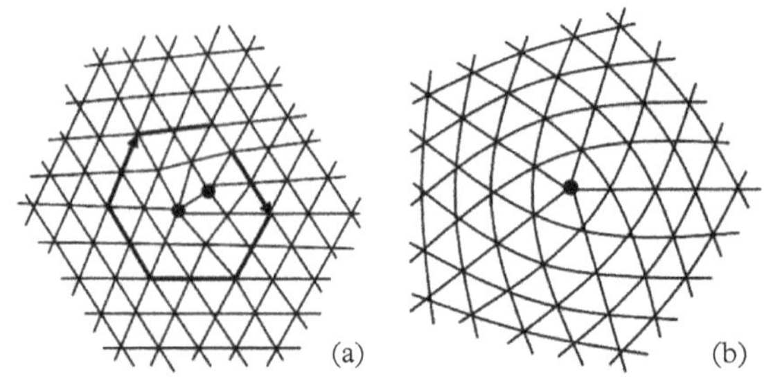

Figure 3.15 *Kosterlitz-Thouless process for two-step melting in two dimensions. (a) Thermal generation of disclination pairs. The marks indicate a fivefold and a sevenfold coordination. (b) Unbinding of the pairs leading to a liquid phase.*

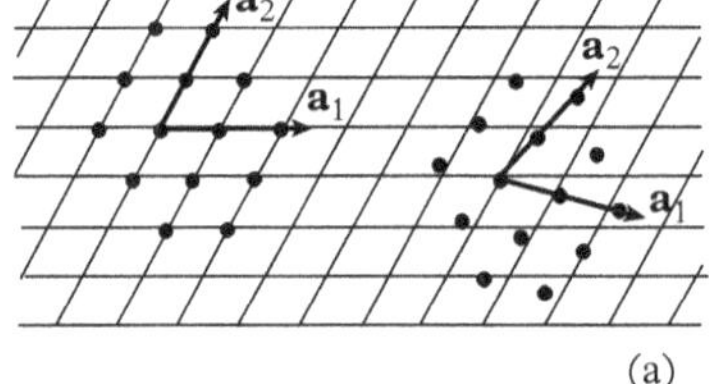
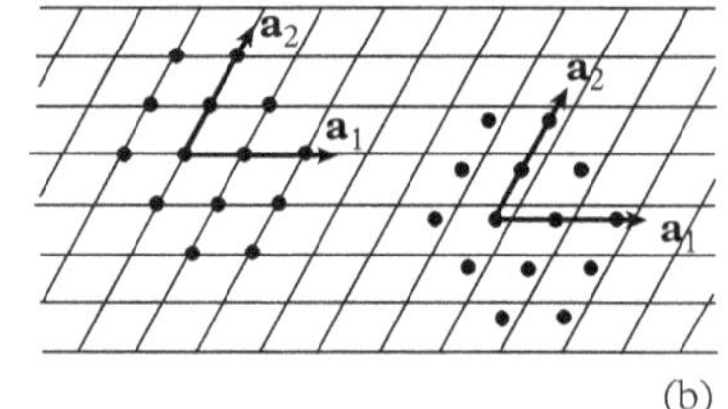

Figure 3.16 *(a) Liquid state in two dimensions illustrated by two separated cluster clusters of short-range order without mutual correlation. (b) Hexatic order: The positions within the two clusters are still uncorrelated but the directions of the local lattice vectors are parallel.*

In 1973 Kosterlitz and Thouless gave a general mechanism of phase transitions in two dimensions as a two-step process. These ideas were extended to the melting of a two-dimensional triangular crystal.[10] As a result, an intermediate phase ('hexatic' phase) occurs between crystalline solid and isotropic liquid. The process is illustrated in Figure 3.15. In the first step of melting, pairs of disclinations are generated thermally. In such a pair the local surroundings of the two 'atoms' change from sixfold to fivefold and sevenfold, respectively (see Figure 3.15a). This leads to destruction of long-range positional order but preserves the directions in the lattice. Only upon further unbinding of these disclination pairs does a liquid phase result (Figure 3.15b).

The intermediate hexatic phase possesses so-called bond-orientational or hexatic order.[11] The directions of the lattice vectors remain ordered (retaining hexagonal symmetry) while the lattice positions possess only short range order (local clusters with a correlation length ξ). This distinction is illustrated in Figure 3.16. In x-ray scattering we can expect a hexatic phase to show broad liquid-like peaks, but a uniform sample will still show maxima indicating hexagonal symmetry. Examples are found in several systems, prominently in liquid crystals. The transition from a CrB phase (crystallized layers) to SmA (liquid layers) indeed may follow a two-stage process with a SmB phase (short for SmB_{hex} these days) in between (Brock et al., 1989). We shall discuss a beautiful example in the case study of Section 3.5.

[10] For more details we refer to the review by Brock et al. (1989).

[11] Bond-orientational order is a somewhat curious term, as no 'bonds' in the conventional sense of the word are involved. Hence we shall use systematically hexatic order.

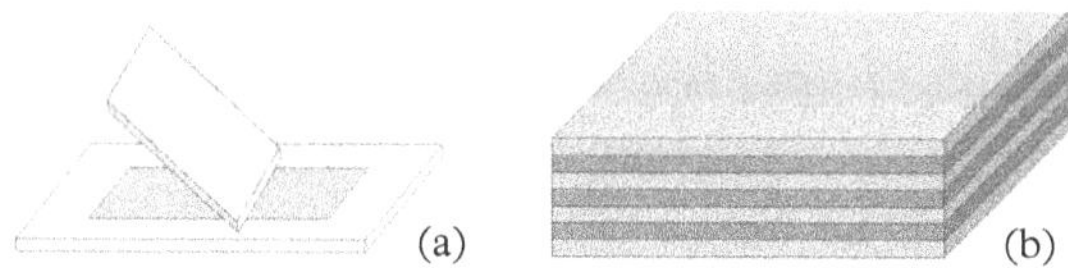

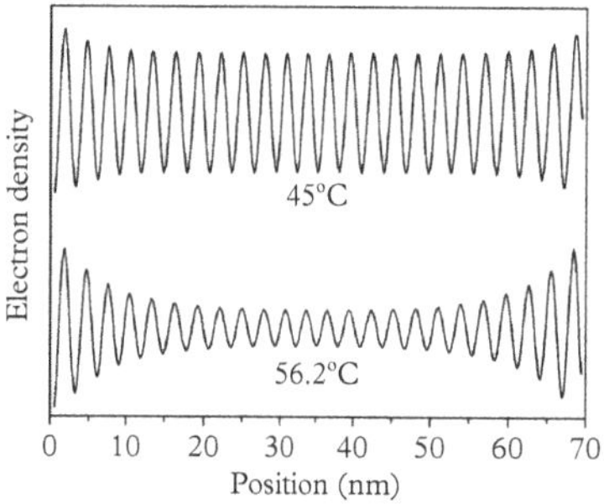

Figure 3.17 *Smectic membrane. (a) Preparation. (b) Resulting structure.*

3.5 Case study: Order in smectic membranes

A unique property of smectic liquid crystals is that, due to the layered structure, films can be freely suspended over an aperture in a frame (smectic membranes).[12] They can easily be made by moving a spreader with smectic material over a hole with sharp edges in a flat substrate (see Figure 3.17). In such films, the smectic layers align parallel to the two air-film interfaces, which are flat because the surface tension minimizes the surface energy of the film.[13] Apart from the edges, such films can be considered as substrate-free and they are up-down centro-symmetric. They have a high degree of uniformity: the alignment of the smectic layers is almost perfect. The remaining mosaic distribution (1 up to 10 mdeg) is mainly due to the non-planarity of the supporting frame edges. Because of the excellent stability of the films (weeks or more), this allows us to study single-domain samples of various thicknesses. The surface area can be large; typical sizes are 10×70 mm^2 or 50 mm Ø. The thickness can be varied from thousands of layers (tens of μm) down to two layers (about 5 nm). Single layers are difficult (less stable) but not impossible. Membranes thicker than some hundred layers can be considered as three-dimensional bulk systems while thin membranes approach two-dimensional behaviour. In the present context, the latter aspect is of special interest.

The first point to be discussed is the observation of stabilization of the smectic layers at the surfaces. This is illustrated in Figure 3.18 for the compound 7AB (see Figure 3.5). In the 24-layer film presented, the amplitude of the density profile increases towards the film-air interfaces. This effect is particularly dramatic if the temperature is increased and the transition to the nematic phase is approached. When the amplitude in the middle of the films approaches zero, the transition into the nematic phase is reached and in principle the film ruptures. However, as a consequence of the surface order, the smectic-A to nematic transition temperature T_{AN} increases for thinner films. For the 24-layer film, $T_{\mathrm{AN}}(24)$ is about 3°C higher than the bulk value $T_{\mathrm{AN}}(\infty) = 53.5$°C.

An even more dramatic effect is that free surfaces may stabilize a higher-ordered phase that is only observed at lower temperatures in the bulk or not observed in the bulk at all. A nice case is illustrated in Figure 3.19 for a seven-layer smectic membrane of 4O.8. As indicated in Figure 3.5, this compound shows in bulk upon heating a transition CrB to SmA. In the seven-layer film this transition is modified near the surfaces. Upon heating, the system melts layer by layer via a two-step process via an intermediated hexatic phase (to be indicated as SmB). Upon cooling, a similar crystallization sequence is found. The phase diagram of

Figure 3.18 *Density profiles of a 24-layer smectic membrane of 7AB upon approaching the smectic–nematic transition at 56.3°C.[14] In bulk this transition is at 53.5°C (after Mol et al., 1998).*

[12] Alternative names are 'free-standing (or freely-suspended) smectic films'.

[13] These films have been known for a long time. Friedel used them in his classical monograph on liquid crystals (1922) as an argument in favour of the existence of layers in the smectic phase. For a review of more recent work on smectic membranes, see de Jeu, Ostrovskii, and Shalaginov (2003).

[14] These profiles have been obtained from x-ray reflectivity, discussed in Chapter 6.

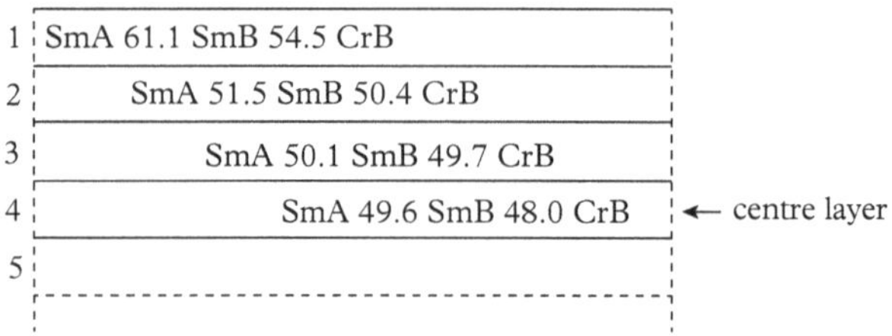

Figure 3.19 *Phase transitions (°C) of the individual layers in a seven-layer smectic membrane of 4.O8 (after Fera et al., 1999). The bulk SmA–CrB transition is at 48.5°C.*

Figure 3.19 indicates transition temperatures for the outermost layers that differ from the next ones.

Most conveniently for experiments, the outermost layers complete their SmA–SmB–CrB freezing transitions before the second outermost layers start their own series. This allows us to follow the transitions in the single top layers on a 'substrate' of the inner liquid layers. In transmission experiments the beam is perpendicular to the smectic layers which thus cannot be observed. However, we get a full view of the relevant changes in the in-plane structure. Such measurements have been taken with electron diffraction and are reproduced in Figure 3.20.

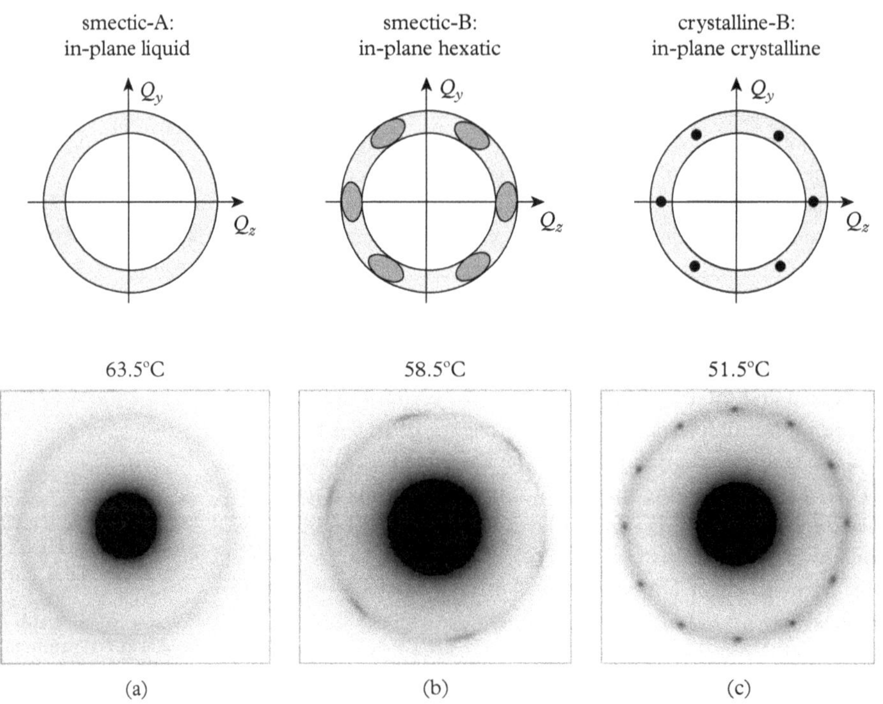

Figure 3.20 *Transmission electron diffraction of a seven-layer 4O.8 membrane indicating a two-step crystallization process of the outermost layers. (Adapted with permission from Chao et al., 1996. Copyright American Physical Society.) (a), (b) and (c) indicate the successive phases of the outermost layers as indicated on top.*

In Figure 3.20a the film is in the high-temperature SmA phase and we observe a diffuse ring from the in-plane liquid structure of all seven layers. On lowering the temperature (Figure 3.20b), we still see the diffuse ring from the five inner layers but in addition the scattering peaks from the two outermost hexatic structures. As mentioned earlier, these should still be broad and show sixfold symmetry. Thanks to the perfectness of the films, this is clearly confirmed while the peaks of the two outer layers are in register. Upon further cooling, the two outermost layers crystallize and the peaks become narrow (Figure 3.20 c). The crystallization of the upper and lower top layer is not in register anymore and we obtain 12 peaks instead of the anticipated six ones. In conclusion, we observe melting/crystallization as a two-step process with an intermediate hexatic phase. This interpretation has been confirmed by various other techniques like calorimetry and the response to shear.

4

Diffraction Physics: Scattering by Crystals

In this chapter we consider diffraction by crystals, the heart of classical crystallography. This is a vast topic and we have to make specific choices, keeping applications to soft matter in mind. The first section summarizes basic elements of crystallography, without going into detailed symmetry considerations. The heart of the chapter is a discussion of diffraction by a crystal lattice, which includes a further discussion of the Ewald sphere and the attendant concept of reciprocal space. After treating some practical aspects of scattering by a crystal, we finish with a case study of polymer crystallization.

4.1 Elements of crystallography

4.1.1 Different types of lattice

A crystal can be considered as a repetitive system (lattice) which possesses certain symmetry elements. The principle is nicely illustrated by the citation in Box 4.1. The repeated structural entity is called the unit cell. For any given lattice the choice of the unit cell is not unique. Figure 4.1 shows possible choices for a general two-dimensional oblique lattice. Note that there is not a single unique oblique

Box 4.1. Citation from W. L. Bragg (1965)

In a two-dimensional design, such as that of a wall-paper, a unit of pattern is repeated at regular intervals. Let us chose some representative point in the unit of pattern, and mark the position of similar points in all the other units. If these points be considered alone, it will be seen that they form a regular network. By drawing lines through them, the area can be divided into a series of cells each of which contains a unit of the pattern. It is immaterial which point of the design is chosen as representative, for a similar network of points will always be obtained.

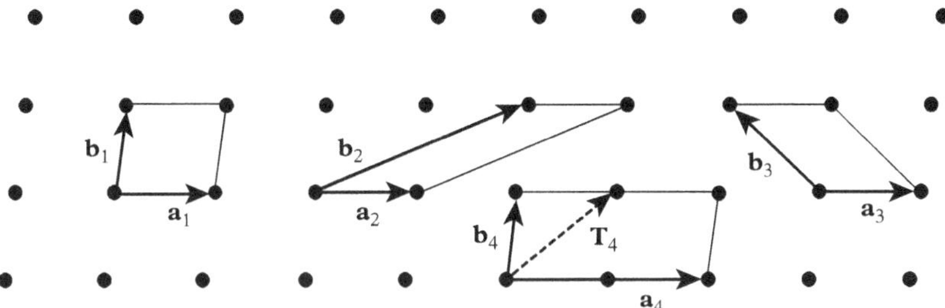

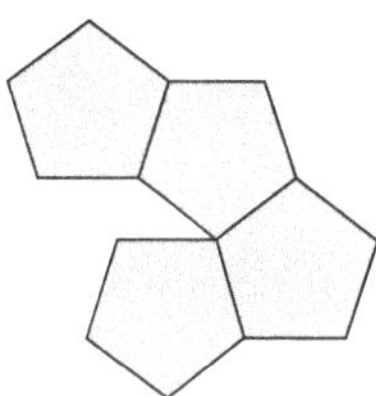

Figure 4.1 *General oblique lattices in two dimensions.*

lattice; the actual situation depends on the length of the primitive vectors. The unit cell with the smallest possible volume is called the primitive (unit) cell. A translation vector $\mathbf{T} = \mathbf{R}_n - \mathbf{R}_{n'}$ connects equivalent points $\mathbf{R}_n = n_1\mathbf{a} + n_2\mathbf{b}$. As in the translation $\mathbf{T}_4$ (in Figure 4.1) cannot be formed by an integral combination of $\mathbf{a}_4$ and $\mathbf{b}_4$, the latter are not primitive translation vectors.

The number of unique lattices is restricted by imposing symmetry elements to the translations. In two dimensions, symmetry operations compatible with translations are mirror, glide, and two-fold, three-fold, four-fold, and six-fold rotations. It can easily be seen that, for example, a five-fold axis cannot be a symmetry element in two dimensions (Figure 4.2).[1] Applying these principles, we can obtain the two-dimensional crystal systems displayed in Figure 4.3.

Figure 4.2 *A plane cannot be fully covered using objects with five-fold symmetry.*

The same principles as illustrated in two dimensions can be applied to three-dimensional structures. Equivalent points in unit cells in a three-dimensional lattice form a periodic structure. Any lattice point can be specified by an integral linear combination of independent primitive translation vectors,[2]

$$\mathbf{R}_n = n_1\mathbf{a} + n_2\mathbf{b} + n_3\mathbf{c}. \tag{4.1}$$

This leads to the 14 different (Bravais) lattices, as summarized in Table 4.1 and Figure 4.4. In turn, the internal structure of the unit cell can be expressed using specific symmetry elements. In combination with the Bravais lattices this creates 230 possible arrangements: the so-called space groups.[3]

[1] A fuller discussion of symmetry elements is beyond the scope of this book.

[2] Alternatively, we shall also use $\mathbf{R}_n = n_1\mathbf{a}_1 + n_2\mathbf{a}_2 + n_3\mathbf{a}_3$.

[3] For a more complete discussion see, for example, http://www.xtal.iqfr.csic.es/Cristalografia/index-en.html.

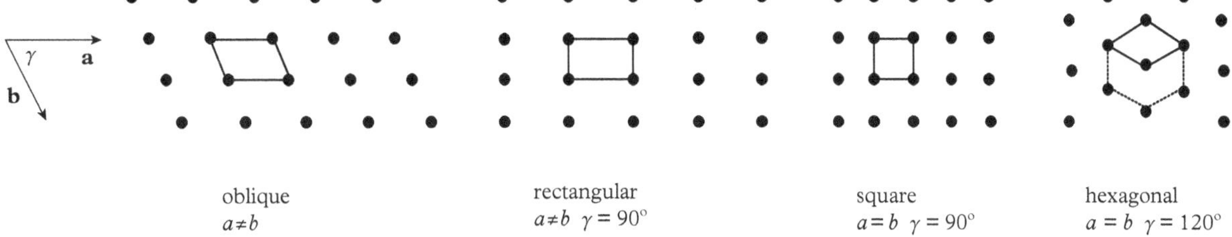

Figure 4.3 *The four unique crystal systems in two dimensions. The lattice of the rectangular system can be either primitive or centred.*

Table 4.1 *The seven crystal systems and 14 Bravais lattices in three dimensions.*

Lattice system	Lattice type	Axes	Axial angles
Triclinic	Primitive (P)	$a \neq b \neq c$	$\alpha \neq \beta \neq \gamma \neq 90°$
Monoclinic	Primitive Base-centred (C)	$a \neq b \neq c$	$\alpha = \gamma = 90° \neq \beta$
Orthorombic	Primitive Body-centred (I) Face-centred (F) Base-centred	$a \neq b \neq c$	$\alpha = \beta = \gamma = 90°$
Tetragonal	Primitive Body-centred	$a = b \neq c$	$\alpha = \beta = \gamma = 90°$
Hexagonal	Primitive	$a = b \neq c$	$\alpha = \beta = 90°$ $\gamma = 120°$
Trigonal	Rhombohedral (R)	$a = b = c$	$\alpha = \beta = \gamma \neq 90°$
Cubic	Primitive Body-centred Face-centred	$a = b = c$	$\alpha = \beta = \gamma = 90°$

4.1.2 Lattice planes and Miller indices

A specific lattice plane is described by the so-called Miller indices (*hkl*) to be constructed as follows.[4]

- Extend the plane to make it cut the crystal axis system at points (a_1, b_1, c_1). If the plane happens to passes through the origin, translate the unit cell in a suitable direction.

- Calculate the reciprocals of the intercepts: $(1/a_1, 1/b_1, 1/c_1)$. If a plane is parallel to an axis, we assume it cuts at infinity and take $1/\infty = 0$.

- Multiply or divide by the highest common factor to obtain the smallest integer numbers (*hkl*).

The process is illustrated in Figure 4.5, leading for this particular case to the triplet (233). Inversely, a plane indexed as say (421), intercepts the **a**-axis at 1/4, the **b**-axis at 1/2, and the **c**-axis at 1/1. Evidently one can find a large number of lattice planes parallel to the one chosen. The agreement is to use round brackets () to describe a single plane and curly brackets {} to describe a family of parallel planes. Thus another plane from the series {421} could intercept the **a**-axis at 1, the **b**-axis at 2, and the **c**-axis at 4. In Figure 4.6, some important planes in a cubic

<hr>

[4] For nice animations explaining Miller indices and lattice planes, see http://www.doitpoms.ac.uk/tlplib/miller_indices.

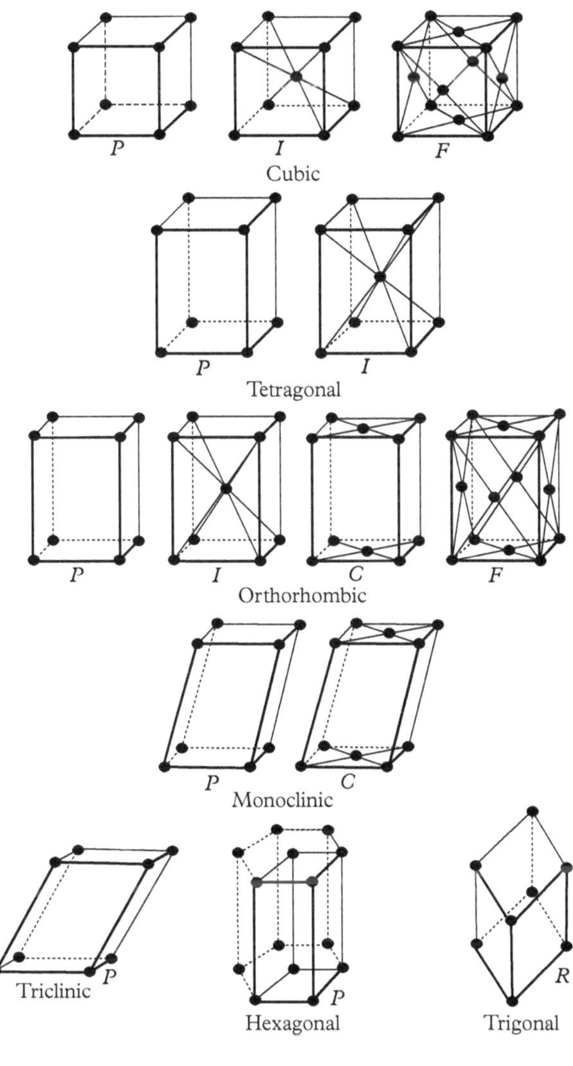

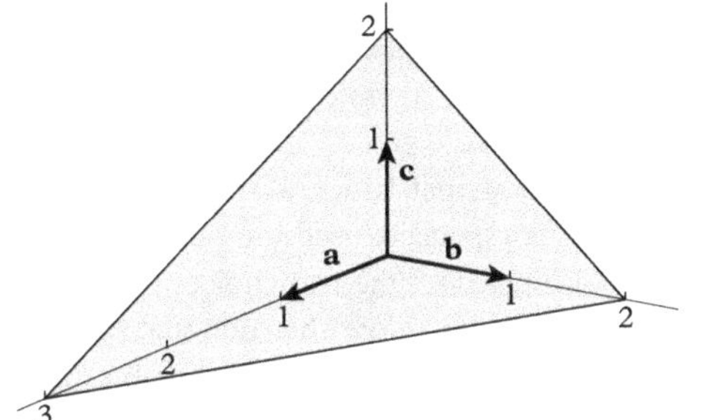

Figure 4.4 *The 14 Bravais lattices in three dimensions (compare with Table 4.1).*

Figure 4.5 *The plane intercepts the* **a**, **b**, *and* **c** *axes at 3a, 2b, and 2 c. The reciprocals of the coefficients are 1/3, 1/2, and 1/2 and the smallest three integers with the same ratio are 2, 3, and 3. Hence the Miller indices (hkl) of this plane are (233) while the whole series of parallel planes is indicated by {233}.*

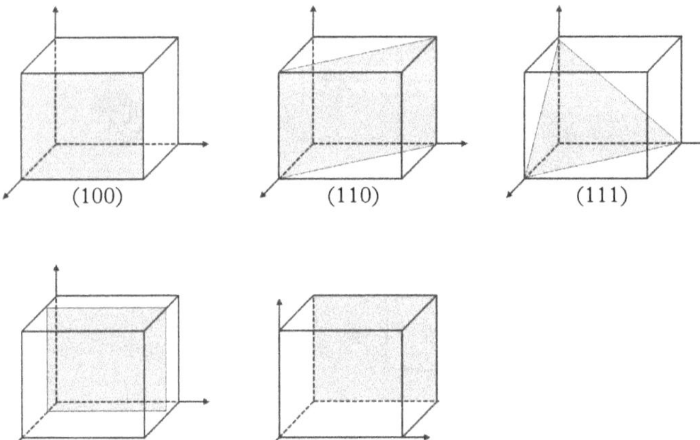

Figure 4.6 *Miller indices of some important planes in a cubic crystal. The planes (300) and ($\bar{1}$00) are parallel to (100).*

Table 4.2 *Perpendicular distances between planes {hkl} of some common lattices.*

Lattice	$1/d^2_{hkl}$	Unit volume V
Cubic	$\dfrac{1}{a^2}(h^2 + k^2 + l^2)$	a^3
Orthorhombic	$\dfrac{h^2}{a^2} + \dfrac{k^2}{b^2} + \dfrac{l^2}{c^2}$	abc
Hexagonal	$\dfrac{4}{3}\left(\dfrac{h^2 + hk + k^2}{a^2}\right) + \dfrac{l^2}{c^2}$	$\dfrac{1}{2}a^2 c\sqrt{3}$
Monoclinic	$\dfrac{1}{\sin^2\gamma}\left(\dfrac{h^2}{a^2} + \dfrac{k^2}{b^2} + \dfrac{l^2\sin^2\gamma}{c^2} - \dfrac{2hk\cos\gamma}{ab}\right)$	$abc\sin\gamma$

crystal are shown together with their indices. In the case of negative integers, the minus sign is replaced by a bar over the number, i.e. $-h$ is written as $\bar{h}$.

Finally, the interplanar spacing d_{hkl} (or the perpendicular distance between neighbouring planes) can be calculated for the various types of lattice from the Miller indices and the value of the unit-cell parameters; see Table 4.2. Often it is convenient to index a structure from the ratio of the observed peaks (see Table 5.1 in the next chapter). For example, a hexagonal structure is characterized by $Q/Q_0 = 1 : \sqrt{3} : 2 : \sqrt{7} : 3$ etc.

In some situations a hexagonal lattice is more conveniently indexed on an orthorhombic basis. This is especially useful to determine small deviations from hexagonal symmetry. In Figure 4.7 the two choices of the basis vectors **a** and **b** are illustrated in two dimensions. Note that now the indexing depends on the particular choice of the basis. Using the orthorhombic basis, deviations from hexagonal symmetry turn up in differences of the value of a_{ortho}/b from $\sqrt{3}$.

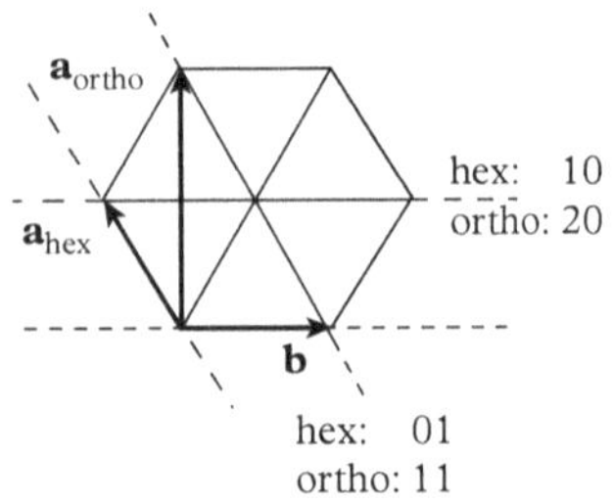

Figure 4.7 *Two ways of indexing of the base plane of a hexagonal lattice. For the original indexing $a_{\mathrm{hex}} = b$; for orthorombic indexing $a_{\mathrm{ortho}} = b\sqrt{3}$.*

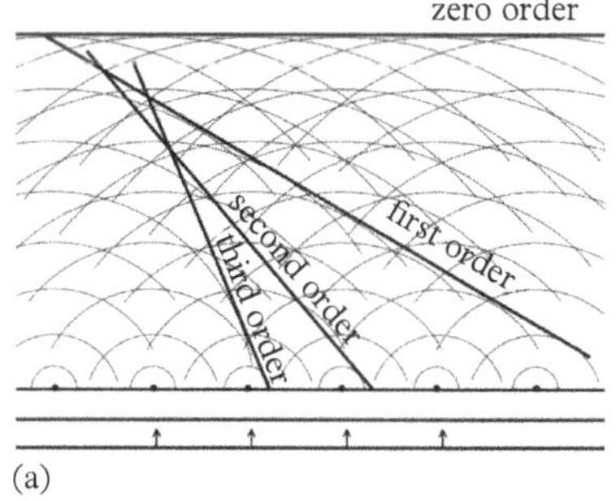

(a)

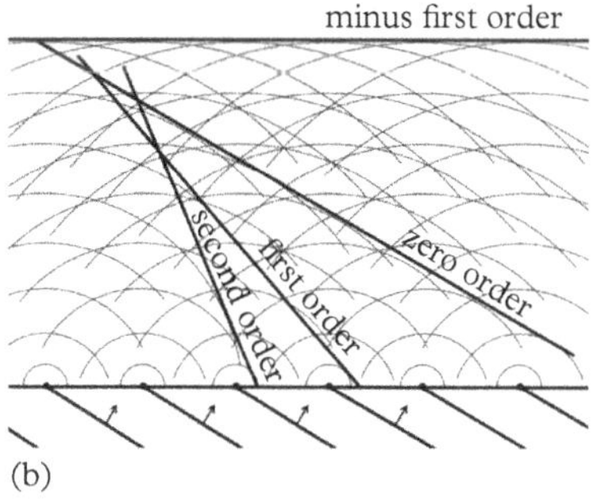

(b)

Figure 4.8 *Interference when a wave front meets a series of atoms in a plane. In (a) the incident wave front is parallel to the lattice plane; in (b) it is at an angle.*

4.2 Diffraction by a crystal lattice

4.2.1 Bragg law

Bragg law is the best known x-ray result everybody seems to know. In many texts a discussion can be found in which a parallel x-ray beam is 'reflected' from a series of lattice planes, and using interference the Bragg formula is 'derived'. The quotation marks are not inserted without purpose, because elementary wave optics teaches us that an electromagnetic wave cannot be reflected from a series of points. Let us consider what is really happening and why the final result is still correct. In Figure 4.8 we consider a one-dimensional row of equally spaced atoms. When an x-ray wave front hits this lattice plane according to Huygens Principle each atom can be considered as a new centre radiating spherical wave shells of x-rays. They form new wave fronts of various orders at angles of incidence for which the scattered waves happen to be in phase.

Considering the various wave fronts, the Bragg formula can be derived in the conventional way by calculating the path difference $2\delta = 2d\sin\theta$ (see Figure 4.9). This leads to the well-known result that constructive interference occurs for

$$2d\sin\theta = k\lambda, \tag{4.2}$$

in which the $k = 1, 2, 3\ldots$ is the order of the diffraction. At this point, the question often arises as to why we need to further discuss x-ray diffraction? Why do we not stop here and stick simply to the application of the Bragg law? In this context, it should be realized that Bragg law only tells us when some scattered intensity might be non-zero. It gives no further information on the intensity $I(Q)$ (that can be zero) and provides no clue as to how structures could be determined.

In spite of the arguments just given, there is one qualitative point regarding intensities that is worthwhile considering at this point. Figure 4.10 demonstrates (for simplicity only in two dimensions) that lines of low Miller indices have the

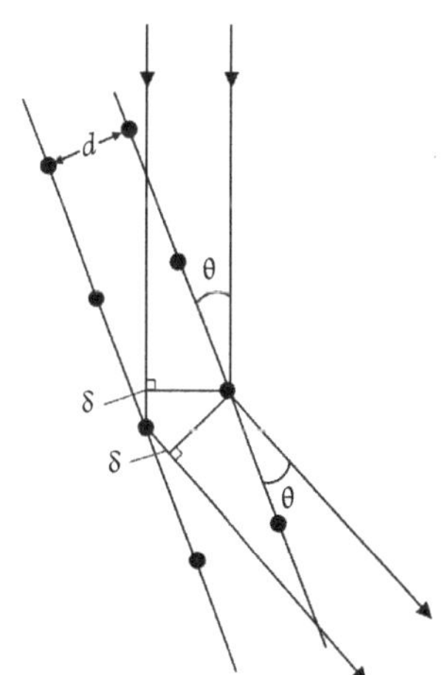

Figure 4.9 *Interference of the zero-order wave front of Figure 4.8 at a series of lattice planes. Constructive interference occurs if the path difference 2δ equals an integer times the wavelength λ.*

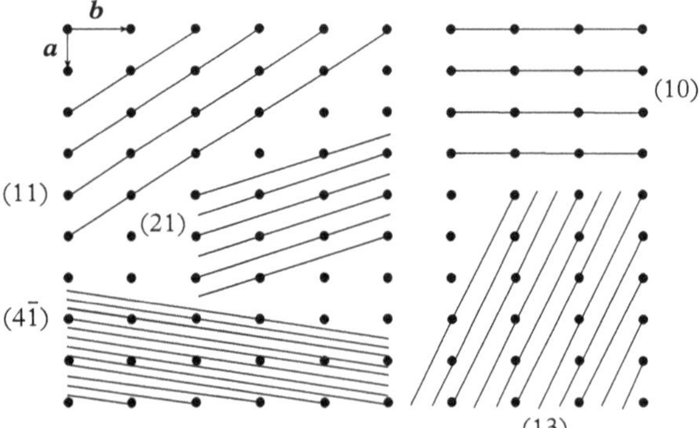

Figure 4.10 *Relations between line index and spacing in a two-dimensional lattice.*

largest spacing and the greatest density of lattice points. Hence, in general, planes of lower indices correspond to stronger scattering. This is the starting point for assigning *hkl*-values to a series of x-ray peaks. More generally, the information from an x-ray peak contains essentially three elements: position (Bragg angle θ or scattering vector Q), intensity, and line shape; these are all needed for a complete structural picture. To be more complete we need the concept of reciprocal lattice (or equivalently reciprocal space), which will be considered in Section 4.2.2.[5]

4.2.2   Introducing reciprocal space

For an overview of the scattering from a crystal lattice one would like to have a handy summary of the lattice planes $\{hkl\}$ and their respective interplanar spacings d_{hkl}. An obvious way would be to represent each plane as a vector of magnitude d_{hkl} directed from the origin of a unit cell perpendicular to the first plane of the $\{hkl\}$ family. We shall illustrate this idea in two dimensions using the lattice planes of Figure 4.10. The result is given in Figure 4.11a. In principle this figure can be extended to four quadrants. However, if we add more planes of higher indices, evidently the sheaf of vectors near the centre will be become very dense. As a result this attempt of an overview is not very useful. Ewald proposed that instead of plotting the vectors corresponding to d_{hkl}, the reciprocal of these vectors should be considered, given by

$$d_{hkl}^* = \frac{1}{d_{hkl}}. \tag{4.3}$$

[5] The term 'reciprocal space' is often experienced as somewhat intimidating. Following largely the treatment of Jenkins and Snyder (1994), we hope to show that there is no need for such a feeling.

Figure 4.11a can now be reconstructed plotting the reciprocal vectors d_{hkl}^*—for the present two-dimensional case d_{hk}^*—as displayed in Figure 4.11b. The units are reciprocal nm (nm⁻¹) and the space is therefore called reciprocal space. Note that the points in this space repeat at perfectly periodic intervals, defining the

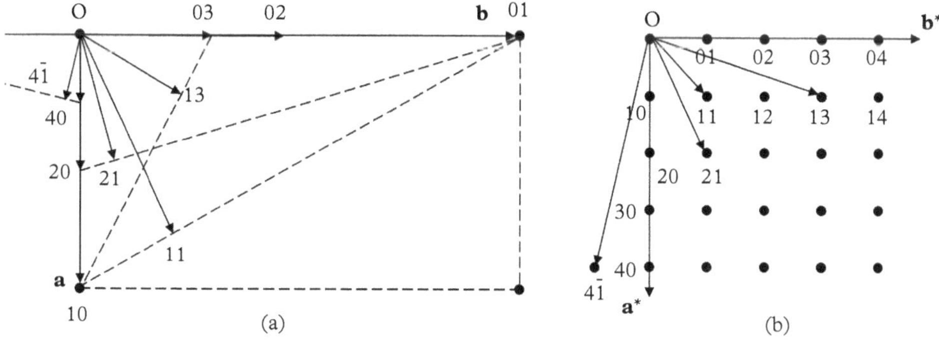

Figure 4.11 *(a) Representation of the two-dimensional lattice planes of Figure 4.10 by vectors d_{hk} in the unit cell. (b) Corresponding two-dimensional reciprocal lattice by plotting $d_{hk}^* = d_{hk}^{-1}$.*

reciprocal lattice. The repeating translation vectors in this lattice are called $\mathbf{a}^*$, $\mathbf{b}^*$, and $\mathbf{c}^*$. Any vector in the reciprocal lattice represents a set of Bragg planes and can be resolved into its components

$$d_{hkl}^* = h\mathbf{a}^* + k\mathbf{b}^* + l\mathbf{c}^*. \qquad (4.4)^6$$

The reciprocal lattice makes the visualization of Bragg planes of atoms very easy by representing them by reciprocal points. Figure 4.11b shows the *(hk)* planes of our two-dimensional lattice of Figure 4.10, which can be considered as the *(hk0)* planes of a fully three-dimensional reciprocal lattice. When connected, the innermost points in the reciprocal lattice will define a three-dimensional shape that is directly related to the shape of the real-space unit cell. Thus the symmetry of the real-space lattice propagates into the reciprocal lattice. In Box 4.2 a more formal definition of the reciprocal lattice vectors is given.

Box 4.2. Formal definition of the reciprocal lattice

For a more formal definition of the reciprocal basis vectors the direct-space basis vectors are written as $\mathbf{a}_i$, $i = 1, 2, 3$, instead of $\mathbf{a}$, $\mathbf{b}$, $\mathbf{c}$. The reciprocal basis vectors then must be chosen such that $\mathbf{a}_i \cdot \mathbf{a}_j^* = 2\pi\delta_{ij}$, in which the Kronecker delta δ_{ij} has the value 1 if $i = j$ and equals 0 otherwise. This condition suffices to determine a unique set of reciprocal basis vectors $(\mathbf{a}_1^*, \mathbf{a}_2^*, \mathbf{a}_3^*)$

$$\mathbf{a}_1^* = \frac{2\pi}{v}\mathbf{a}_2 \times \mathbf{a}_3, \quad \mathbf{a}_2^* = \frac{2\pi}{v}\mathbf{a}_3 \times \mathbf{a}_1, \quad \mathbf{a}_3^* = \frac{2\pi}{v}\mathbf{a}_1 \times \mathbf{a}_2,$$

in which $v = \mathbf{a}_1 \cdot (\mathbf{a}_2 \times \mathbf{a}_3) = \mathbf{a}_2 \cdot (\mathbf{a}_3 \times \mathbf{a}_1) = \mathbf{a}_3 \cdot (\mathbf{a}_1 \times \mathbf{a}_2)$ is the volume of the reciprocal unit cell.

[6] Equation (4.4) provides an alternative definition of the Miller indices *hkl* that is, of course, equivalent to the one introduced in Section 4.1.2.

In orthogonal crystal systems, the relationship between **d** and **d*** is simply reciprocal. In non-orthogonal systems (hexagonal, monoclinic, and triclinic) the vector character of the reciprocals complicates the angular calculations. The reciprocal angles α^*, β^*, and γ^* are defined as 180° minus the real-space angle.[7] For orthogonal systems the angular relations are quite simple. For non-orthogonal systems like hexagonal and monoclinic, they are somewhat more complex (see Table 4.3). We refrain from writing out the elaborate result for the triclinic system.

Figure 4.12 illustrates the concept of the reciprocal lattice once more by some examples in one and two dimensions. The one-dimensional example is obviously simple. For a square lattice of dimension a it is readily apparent that the reciprocal lattice is also square with a spacing $2\pi/a$. If the axes are not orthogonal, as for a two-dimensional hexagonal lattice, the basis vectors in direct and reciprocal space are not necessarily parallel, as indicated in the bottom part of Figure 4.12 (see, also, Table 4.3). In three dimensions things are more intricate and will not be discussed in further detail. We just mention as an example that a face-centred cubic lattice in direct space is connected to a body-centred cubic reciprocal lattice.

The condition for diffraction can be easily determined by combining the reciprocal lattice with the Ewald sphere.[8] Once more, we illustrate the principle in two dimensions. Figure 4.13a shows a two-dimensional reciprocal lattice with a

[7] Also called the supplementary angle.

[8] Introduced in Section 2.1.2.

Table **4.3** *Relations between direct and reciprocal space.*

System	$\mathbf{a}^* \equiv \mathbf{a}_1^*$	$\mathbf{b}^* \equiv \mathbf{a}_2^*$	$\mathbf{c}^* \equiv \mathbf{a}_3^*$
Orthogonal	$\mathbf{a}^* = 1/\mathbf{a}$	$\mathbf{b}^* = 1/\mathbf{b}$	$\mathbf{c}^* = 1/\mathbf{c}$
Hexagonal	$\mathbf{a}^* = 1/(\mathbf{a}\sin\gamma)$	$\mathbf{b}^* = 1/(\mathbf{b}\sin\gamma)$	$\mathbf{c}^* = 1/\mathbf{c}$
Monoclinic	$\mathbf{a}^* = 1/(\mathbf{a}\sin\beta)$	$\mathbf{b}^* = 1/\mathbf{b}$	$\mathbf{c}^* = 1/(\mathbf{c}\sin\beta)$

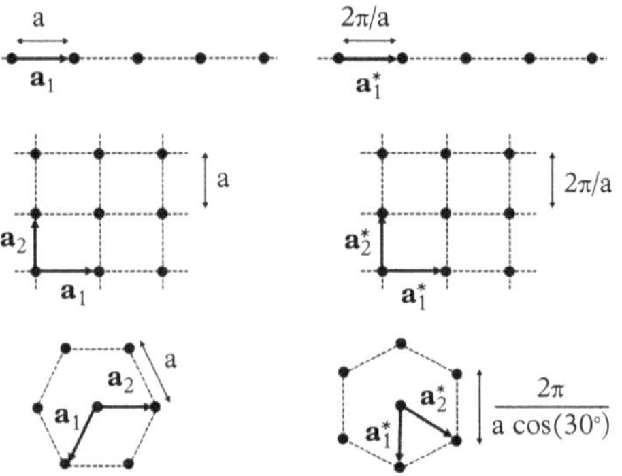

Figure 4.12 *Illustration of real (left) and reciprocal space (right) in one and two dimensions.*

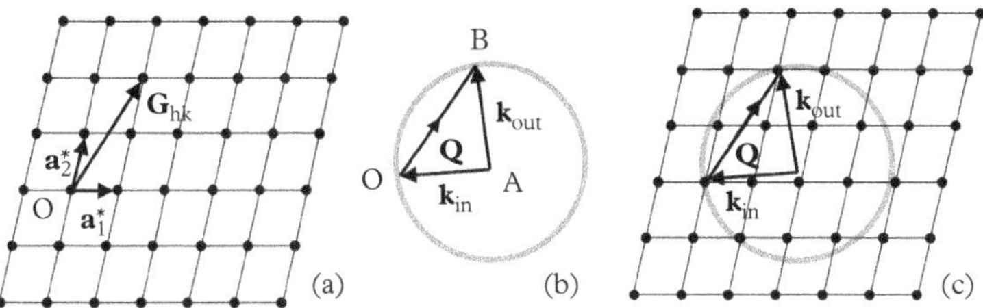

Figure 4.13 *Reciprocal lattice in two dimensions (a) and the Ewald circle (b) are combined in (c) to demonstrate the scattering condition (after Als-Nielsen and McMorrow, 2011).*

particular lattice vector $\mathbf{G} = h\mathbf{a}_1 + k\mathbf{a}_2$. The scattering triangle in Figure 4.13b indicates an incoming x-ray beam along $\mathbf{k}_{in} = \mathrm{AO}$ that can be scattered to any wave vector $\mathbf{k}_{out} = \mathrm{AB}$ terminating on the Ewald circle of radius k. Figure 4.13c is a superposition of (a) and (b) using the common origin O, indicating that scattering occurs if a reciprocal lattice vector terminates on the Ewald circle. Evidently the scattering condition can be written as

$$\mathbf{Q} = \mathbf{G}. \tag{4.5}$$

As we can see, the combined concepts of reciprocal lattice and Ewald sphere (in two dimensions Ewald circle) allow us to explain diffraction in a pictorial model that avoids the need to consider complex crystallographic relationships.

4.2.3   Lattice sum and scattering conditions

Let $\mathbf{R}_n$ be the vectors defining the lattice and $\mathbf{r}_j$ that of the atoms in the unit cell whose lattice site is at the origin (see Figure 2.2b of Chapter 2). Hence the position of any given atom in the sample can be described by $\mathbf{R}_n + \mathbf{r}_j$. We return to the generalized Equation (2.3)

$$F^{\mathrm{crystal}}(\mathbf{Q}) = \sum_{j}^{\mathrm{unit\ cell}} F_j^{\mathrm{atom}}(\mathbf{Q})\exp(i\mathbf{Q}\cdot\mathbf{r}_j)\sum_{n}\exp(i\mathbf{Q}\cdot\mathbf{R}_n) = F^{\mathrm{unit\ cell}}(\mathbf{Q})\sum_{n}\exp(i\mathbf{Q}\cdot\mathbf{R}_n).$$

$$\tag{4.6}$$

The first summation is the *unit cell structure factor*. It is the equivalent of the form factor we encountered earlier for scattering from isolated particles. However, now a summation is involved over the form factors of the atoms or molecules in the unit cell with their respective phase factors. The second term is the *lattice sum* or, alternatively, the structure factor or interference function $S(\mathbf{Q})$. Hence, using $S(\mathbf{Q}) = \sum_{n}\exp(i\mathbf{Q}\cdot\mathbf{R}_n)$ we can write Equation (4.6) as

$$F^{\mathrm{crystal}}(\mathbf{Q}) = F^{\mathrm{unit\ cell}}(\mathbf{Q})S(\mathbf{Q}).$$

Note that the scattering now originates from discrete points. This results in summations instead of integrations as before, due to a continuous function $\rho(\mathbf{r})$.

In Chapter 2 we related the structure factor to the pair distribution function $g_2(\mathbf{r})$ via

$$S(\mathbf{Q}) - 1 = \int \exp(i\mathbf{Q} \cdot \mathbf{r})\,[g_2(\mathbf{r}) - 1]\,d\mathbf{r}.$$

For a lattice the function $g_2(\mathbf{r})$ is discrete and can be written as

$$g_2(\mathbf{r}) = \sum_n \delta(\mathbf{r} - \mathbf{R}_n).$$

Though the lattice sum seems to be a logical step in the sequence point scatterer $\rightarrow$ atom $\rightarrow$ molecule $\rightarrow$ lattice, it is qualitatively very different. Real systems involve a summation over a full crystal, which for $1\,\mu$m size is already about 10^4 times the length of a typical basis vector. In three dimensions this corresponds to about 10^{12} unit cells. Each of the terms corresponds to a phase factor. The summation will only reach a finite value for phases that are multiples of 2π. Any deviation from this value will lead to destructive interference between terms at n, $n + 1, \ldots$ and terms in antiphase at $n + m$, $n + 1 + m$. This is very similar to the situation of light scattering from an optical grating. As a result, to find any diffracted intensity a scattering vector $\mathbf{Q}$ must fulfil the condition:

$$\mathbf{Q} \cdot \mathbf{R}_n = 2\pi \times \text{integer}. \tag{4.7}$$

Let the lattice vectors be given by $\mathbf{R}_n = n_1\mathbf{a}_1 + n_2\mathbf{a}_2 + n_3\mathbf{a}_3$. Hence to obey Equation (4.7) we have to look for the appropriate values of our scattering vector $\mathbf{Q}$. To do so we consider the vectors $\mathbf{G}(\mathbf{Q})$ of the reciprocal lattice $\mathbf{G} = h\mathbf{a}_1^* + k\mathbf{a}_2^* + l\mathbf{a}_3^*$, with basis vectors $(\mathbf{a}_1^*, \mathbf{a}_2^*, \mathbf{a}_3^*)$. From the construction of the Ewald sphere in the previous section we know that the scattering condition Equation (4.7) is fulfilled if $\mathbf{Q}$ equals a reciprocal lattice vector $\mathbf{G}$. In Box 4.3 this condition is considered in a general way; necessarily more formal. Comparing the final result in Box 4.3 with Equation (4.7) leads once more to the diffraction condition (Equation 4.5). It is called the von Laue condition for diffraction and was nicely visualized in Section 4.2.2. The equivalence of the Laue condition and the classical Bragg equation is illustrated once more in Figure 4.14.

Box 4.3. Diffraction condition in reciprocal lattice

To check the diffraction condition, Equation (4.7), for the reciprocal lattice we use Box 4.2 to calculate the first term of $\mathbf{G} \cdot \mathbf{R}_n$ as

$$h\mathbf{a}_1^* \cdot n_1\mathbf{a}_1 = h2\pi \frac{\mathbf{a}_2 \times \mathbf{a}_3}{\mathbf{a}_1 \cdot (\mathbf{a}_2 \times \mathbf{a}_3)} \cdot n_1\mathbf{a}_1 = 2\pi h n_1.$$

The other terms follow in a similar way and the total result is given by

$$\mathbf{G} \cdot \mathbf{R}_n = 2\pi(hn_1 + kn_2 + ln_3) = 2\pi \times \text{integer}.$$

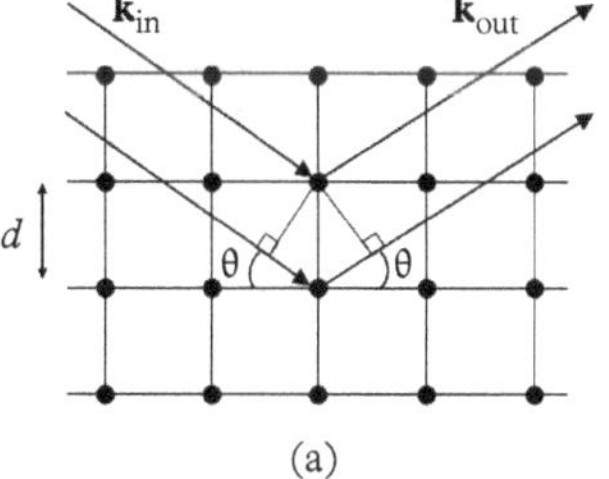
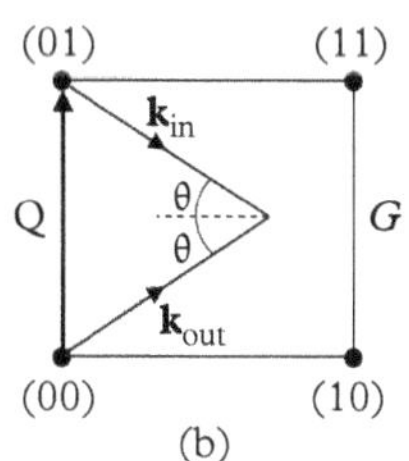

Figure 4.14 *Equivalence of Bragg (a) and Von Laue (b) diffraction conditions (after Als-Nielsen and McMorrow, 2011).*

$$Q = G$$

$$\frac{4\pi}{\lambda}\sin\theta = \frac{2\pi}{d}$$

$$2d\sin\theta = \lambda$$

The arrangement and spacing of certain classes of lattice planes can produce diffractions that are always a factor of π out of phase. This leads to a phenomenon called extinction. As an example we consider a body-centred cubic cell. For each atom located at x, y, and z there will be an identical atom at position $x + 1/2$, $y + 1/2$, and $z + 1/2$. The corresponding structure factor $S(\mathbf{Q}) = S_{hkl} = \sum_j \exp(i\mathbf{Q} \cdot \mathbf{R}_j)$ then can be represented as

$$S_{hkl} = \sum_j \exp\left[2\pi i(hx_j + ky_j + lz_j)\right] + \sum_j \exp\left[2\pi i\left(hx_j + \frac{h}{2} + ky_j + \frac{k}{2} + lz_j + \frac{l}{2}\right)\right].$$

If $h + k + l$ is even, in the second term $(h + k + l)/2$ will be equal to an integer n. An integral number times 2π will have no effect on the value of this term and it will be equal to the first one. Hence the result is

$$S_{hkl} = 2\sum_j \exp[2\pi i(hx_j + ky_j + lz_j)].$$

However, if $h + k + l$ is odd, the second term will contain an integer with a $2\pi\,(n/2)$ term. This introduces a minus sign in the second term and the net result is $S_{hkl} = 0$. There is no diffracted intensity. This condition is called a systematic extinction. The presence or absence of systematic extinctions provides an important clue to the crystal class of a particular system.[9] In Table 4.4 we give the most important extinction conditions for the various lattice types seen in Table 4.1.

At the end of this section two final remarks seem appropriate. First, recall from Section 2.1.3 the discussion of the phase problem that can be adapted to the present situation. Though the diffraction pattern is obtained by taking the Fourier transform of the scattering density, it is not so straightforward to determine the

[9] Note that this type of result cannot be obtained from the Bragg equation.

Table 4.4 *Lattice type and some extinction conditions.*

Lattice type	Extinction condition
P (primitive)	none
C (base centred)	$h+k=$ odd
I (body centred)	$h+k+l=$ odd
F (face centred)	h, k, l mixed even and odd

crystal structure from a diffraction pattern with the inverse Fourier transform. We cannot directly measure the relevant structure factor S_{hkl}, but only the diffracted intensity $I_{hkl} = |S_{hkl}|^2$. Hence, only the amplitude of S_{hkl} is determined and not its phase. One approximate way out is to simply ignore this fact and take the inverse Fourier transform I_{hkl} anyway. This leads to the so-called Patterson function which contains peaks corresponding to the unit cell of the crystal. It can serve as a first guess of the crystal structure to be subsequently refined.

Second, we have implicitly made the important approximation that the scattering of the x-rays is weak, which means that multiple scattering effects are disregarded. This is the domain of the *kinematical diffraction theory* as presented here. In the case of perfect single crystals, multiple-scattering occurs and the resulting description becomes necessarily more complex: *dynamical diffraction theory*. Single-crystal studies allows us to get detailed information about the molecular structure and the electron distribution within the unit cell. In terms of soft matter, this is especially important in the area of protein crystallography, which has developed into a huge specialized field. Obviously, this is beyond the scope of this book.

4.3 Miscellaneous properties of crystal scattering

4.3.1 Debye–Waller factor

In any crystal there are considerable thermal fluctuations around the lattice points. Even though they remain finite, the associated changes in nearest-neighbour spacings can easily differ by 10% at room temperature. Should this not prevent the observation of well-defined x-ray peaks?[10] Consider the scattering from a specific lattice direction $\mathbf{R}_n$ that depends on factors of the type $\exp(i\mathbf{G}\cdot\mathbf{R}_n)$. Let us assume that the atomic position contains a term $\mathbf{u}(t)$ fluctuating in time: $\mathbf{R}(t) = \mathbf{R}_n + \mathbf{u}(t)$. Then the lattice sum contains terms of the type

$$\exp(i\mathbf{G}\cdot\mathbf{R}_n)\langle\exp(i\mathbf{G}\cdot\mathbf{u})\rangle,$$

[10] This question was the subject of heavy debates in the early days of x-ray crystallography. The answer, as discussed here, is due to Debye.

where the brackets indicate a thermal average. Expanding the second exponential[11] gives

$$\langle \exp(i\mathbf{G} \cdot \mathbf{u}) \rangle = 1 + i\langle \mathbf{G} \cdot \mathbf{u} \rangle - \frac{1}{2}\langle (\mathbf{G} \cdot \mathbf{u})^2 \rangle + \ldots .$$

In this situation $\mathbf{u}$ is a random thermal displacement that is not related to the direction of $\mathbf{G}$. As a result we obtain $\langle \mathbf{G} \cdot \mathbf{u} \rangle = 0$. Furthermore, we can write

$$-\frac{1}{2}\langle (\mathbf{G} \cdot \mathbf{u})^2 \rangle = -\frac{1}{2} G^2 \langle u^2 \rangle \langle \cos^2\theta \rangle = -\frac{1}{6}\langle u^2 \rangle G^2,$$

in which the factor 1/3 arises from the average of $\cos^2\theta$ over a sphere. We can consider the remaining factor as the leading term in an expansion of $\exp(-\frac{1}{6}\langle u^2 \rangle G^2)$. Then the original term in the lattice sum becomes

$$\exp\left(i\mathbf{G} \cdot \mathbf{R}_n\right) \exp\left(-\frac{1}{6}\langle u^2 \rangle G^2\right).$$

The corresponding intensity then can be written as the quadratic form

$$I = I_0 \exp\left(-\frac{1}{3}\langle u^2 \rangle G^2\right), \tag{4.8}$$

in which I_0 gives the scattering from the rigid lattice and the second term is called the Debye–Waller factor. As we can see, in the presence of thermal fluctuations any x-ray peak remains well-defined. Only the intensity of the diffracted beam will be reduced by the thermal fluctuations, as given by the term involving the mean-square displacement $\langle u^2 \rangle$.

4.3.2   Mosaic spread

Any oriented sample (crystalline, smectic, etc.) will in general not be perfect. It can be modelled as a series of small uniform blocks (crystallites) that are distributed around the average orientation; see Figure 4.15a.[12] This effect is static, and thus well distinguished from the fluctuations just discussed. The mosaic spread or mosaicity can be measured by rocking the sample while keeping the scattering angle 2θ constant. The result will indicate a distribution of $\mathbf{Q}$ around its average direction (see Figure 4.15b) that is, for a crystal, often small. It can usually be described by a Gaussian. For anisotropic particles the width of this distribution can be used as a rough measure of the degree of alignment. As long as there is no phase relation between the various positions, the intensities from the blocks can be simply added. This assumes that the diffracted beam continues in a specific block without being rescattered (kinematical approximation, see the end of Section 4.2.3).

---

[11] $\exp x = 1 + x + \frac{x^2}{2!} + \frac{x^3}{3!} + \ldots$

[12] The mosaic crystal model goes back to a theoretical analysis of x-ray diffraction by C. G. Darwin.

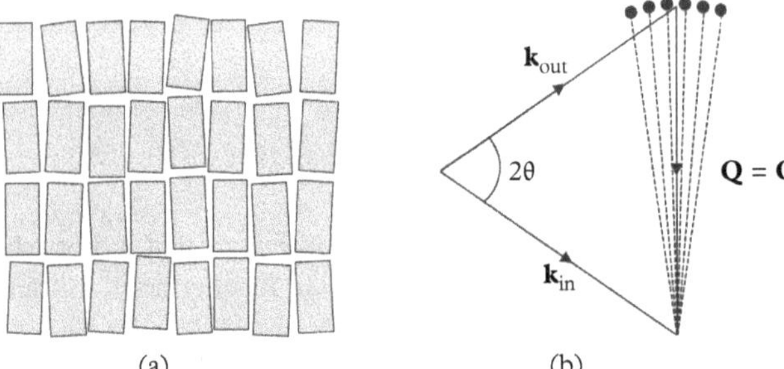

Figure 4.15 *Illustration of the mosaic distribution in a crystalline sample (a) and its effect on the scattering vector (b).*

4.3.3   Line width and average domain size

Apart from its position and intensity, the line shape of an x-ray peak is also highly relevant. In the case of long-range order in crystals, the most important point is that the Fourier transforms involved in the scattering are essentially integrals over an infinite volume. In contrast, real samples have a finite size, which leads to broadening of the scattered intensity.[13] Inversely, information on the average crystal size (or the size of ordered domains) can be obtained from the line width, usually quantified as the full-width-at-half-maximum (FWHM). This quantity can be written as $\Delta Q\,(\mathrm{nm}^{-1})$ and the attendant average domain size is then given by

$$L = 2\pi/\Delta Q\,(\mathrm{nm}). \tag{4.9}$$

In older literature the FWHM is often expressed as $\Delta\theta$ (in radians). Taking the derivative of the expression $Q = (4\pi/\lambda)\sin\theta$ gives $\delta Q = (4\pi/\lambda)\cos\theta \cdot \delta\theta$. As the FWHM is given by $\Delta\theta = 2\delta\theta$, we find a crystal/domain size

$$L = \frac{2\pi}{\Delta Q} = \frac{\lambda}{\Delta\theta \cdot \cos\theta}. \tag{4.10}$$

The latter expression is also known as the Scherrer equation, to which an empirical proportionality factor $K \approx 0.9$ is often added.

In practice, a good measurement of the FWHM is often complicated. The best procedure is to subtract any background from the peak under consideration and fit the result to a Gaussian function. The latter is usually appropriate for long-range order in crystals. The practical procedure will become complicated for multiple overlapping peaks. Furthermore, any experimental x-ray peak is the convolution of the scattering from the sample and the instrumental resolution, which usually can also be described by a Gaussian. The convolution of two Gaussians is again a Gaussian, and the measured experimental width σ_{exp} is given by

---

[13] Compare with Section 2.2.1.

$$\sigma_{\mathrm{exp}}^2 = \sigma_{\mathrm{sample}}^2 + \sigma_{\mathrm{instr}}^2. \tag{4.11}$$

Only if $\sigma^2_{\text{instr}} \ll \sigma^2_{\text{sample}}$ can we equate the FWHM of the experiment with that of the sample. In the other limit we arrive at $\sigma_{\text{exp}} \approx \sigma_{\text{instr}}$ and speak of a resolution-limited peak width.

So far we assumed the crystal under investigation to be perfect. In real crystals occasional substitution of a foreign atom, a vacancy or an interstitial atom, and in the case of polymers a chain end, will create deviations from the ideal structure. The degree of imperfections in crystalline polymers is much higher than in small molecular or ionic crystals. These effects contribute to the line shape, especially at large angles, and complicate the discussion given so far. Two somewhat idealized types of imperfections have been distinguished, pictured in Figure 4.16 for a two-dimensional lattice.[14] Note that Fig 4.16a is the static equivalent of the dynamic effect of the Debye–Waller factor.

4.3.4 Polycrystalline diffraction

A polycrystalline powder consists of many thousands of tiny crystallites with all possible orientations randomly distributed. Reciprocal lattice vectors $\mathbf{G}_{\text{hkl}}$ will have their directions isotropically distributed over the sphere around O' as indicated in Figure 4.17. Grains for which $\mathbf{G}_{\text{hkl}} = \mathbf{Q}$ fulfil the scattering condition.

[14] For a further analysis we refer to Roe (2000).

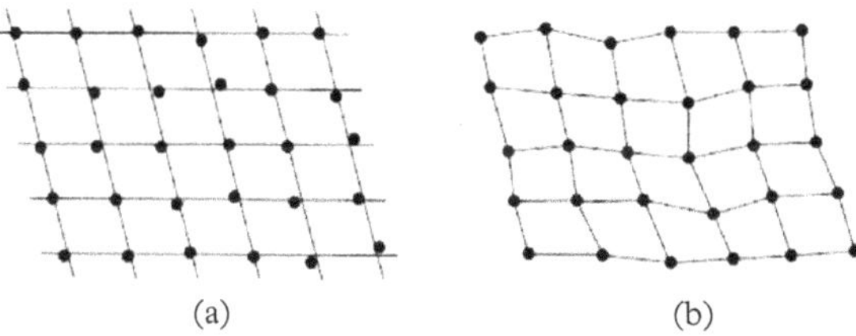

(a) (b)

Figure 4.16 *Crystal imperfections of the first kind (a) and the second kind (b).*

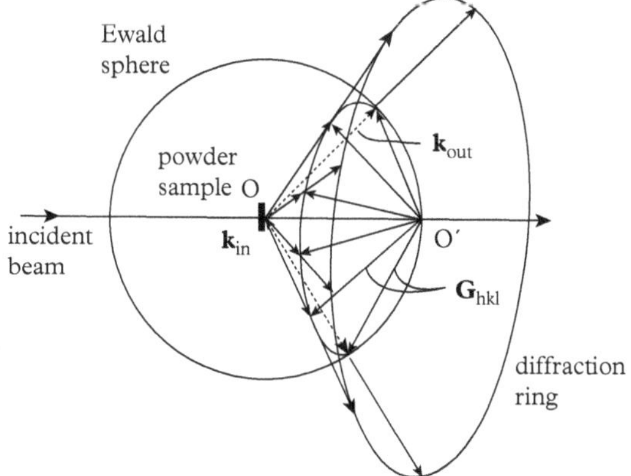

Figure 4.17 *Debye cone for powder diffraction on the Ewald sphere (after Jenkins and Snyder, 1996).*

As a result the scattered wave vectors $\mathbf{k}_{out}$ are distributed evenly on a cone with $\mathbf{k}_{in}$ as the axis. If the crystallites are randomly oriented they produce a continuous diffraction ring (Debye cone, Figure 4.17), a generalization of the two-dimensional case in Figure 2.9. The cone of diffraction patterns consists of numerous closely spaced dots, each from diffraction of a single crystallite. Each ring corresponds to a particular vector $\mathbf{G}$ in the sample.[15]

Figure 4.18a shows powder diffraction data for the polymer polyethylene (PE). We can distinguish two strong lines on a diffuse background and three additional weaker lines at larger angles. After circular averaging we obtain the linear plot of the diffracted intensity I as a function of Q of Figure 4.18b.[16] Alternatively, I has often been plotted against the scattering angle 2θ which, however, makes the diffractogram dependent on the wavelength λ. This makes a comparison of data at different wavelengths (as often happens at modern synchrotron sources) difficult, and we recommend always using Q. This has the additional advantage that on a linear scale ratios of the positions of x-ray peaks show up directly, without the need to correct for the variation of the sin-function in the Bragg equation.

Powder diffraction allows for rapid, non-destructive analysis of unknown materials, including mixtures. As such, it has found widespread application in many scientific fields, including soft condensed matter. Interpretation of a powder pattern involves three steps. The first one involves finding the size and symmetry of the unit cell so that the reflections can be labelled appropriately (h, k, l). The second step is to extract the measured intensities and to convert them into structure factors. Finally, the structural model is recalculated using the entire diffraction profile. A major challenge of polycrystalline diffraction is to obtain reliable intensities. Another point is the degree of randomness of the powder. Larger, isolated crystals may appear as single spots on a diffraction ring. Figure 4.19 illustrates the effect of an increasing number of crystallites on the final x-ray pattern.

[15] In different words: The orientational averaging causes the three-dimensional reciprocal space to be projected onto a single dimension.

[16] Such a plot is called a diffractogram. The relevant Q-range for scattering from crystals is referred to as wide-angle x-ray scattering (WAXS).

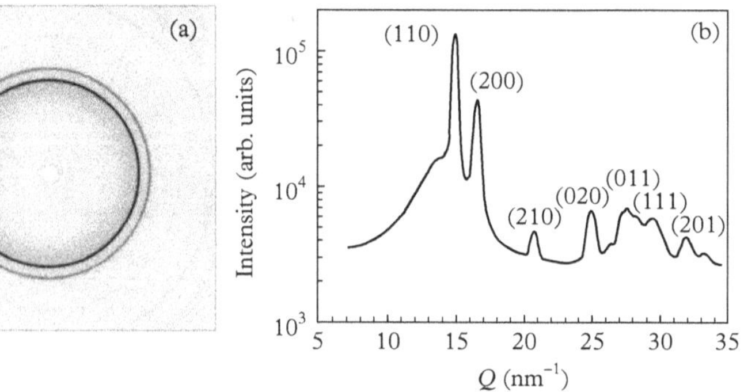

Figure 4.18 *Powder diffraction pattern of PE, $(-CH_2-)_n$. (a) Two-dimensional picture. (b) Diffractogram after circular averaging.*

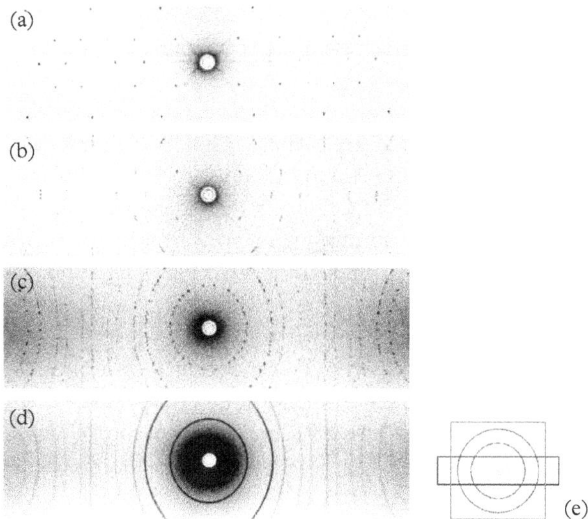

Figure 4.19 *Illustration of the effect of multiple crystal diffraction for fluorite. (a) Single crystal. (b) Single crystal with rotations over 2° with vertical. (c) Five randomly oriented crystals. (d) Powder pattern. (e) Scheme of the section displayed in the full picture. (After Peiser, Rooksby, and Wilson, 1955).*

4.4   Case study: Polymer crystallization

Crystallization of polymers is a huge field with important practical consequences. In industrial extruders materials are made for many rather different applications. In each case the (mechanical) properties depend on the degree of crystallization of the material (amorphous vs. crystalline part) and its distribution. We shall restrict ourselves to a couple of questions in which x-ray scattering plays a crucial role.[17] Let us first consider the experimental picture describing polymer crystallization, which starts with the optical observation of spherulites. On a closer look, the amorphous melt of Figure 4.20a has changed into a semi-crystalline structure of folded chains alternating with amorphous regions (Figure 4.20b). Figure 4.20c gives a model of the semi-crystalline lamellae. Obviously, these structures can be investigated by x-ray scattering. The alternating layer structure is accessible by SAXS. The structure of the crystalline layer itself can be probed with WAXS, mostly using a powder, but in some cases an oriented crystal can also be obtained.

[17] There are numerous reviews of polymer crystallization. A summary of early work is given by Keller (1968). The modern aspects and controversies are well described by Strobl (2006, 2009).

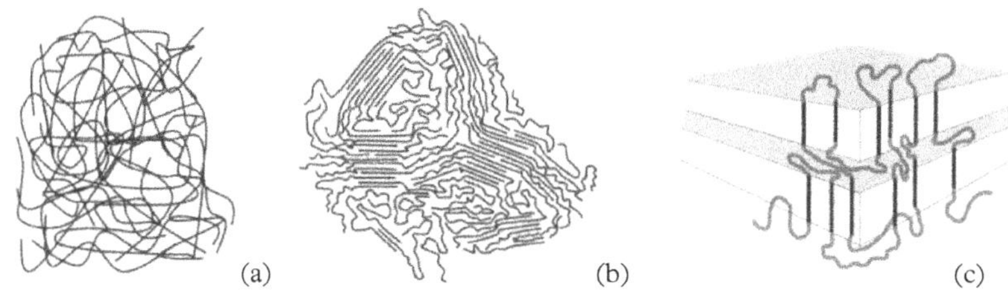

Figure 4.20 *Polymer structure. (a) Amorphous melt. (b) Semi-crystalline: mixed crystalline and amorphous regions. (c) Model of semi-crystalline lamellae.*

The various questions associated with this picture deserve an in-depth discussion, but we shall restrict ourselves to a few comments in the next paragraph. Subsequently some specific experimental x-ray results will be considered.

A good starting point is to consider crystallization of oligomers. Regarding paraffins with uniform chain length, two interesting observations can be noted. Ungar et al. (1985) reported on extended-chain crystals of various lengths and found that above $C_{150}H_{302}$ chain folding sets in. Sirota and coworkers (Kraack, Deutsch, and Sirota, 2000) strongly suggest a crossover from nucleation of entire molecules to nucleation of fractional molecular bundles ('sheaves') occurring around chain lengths as low as n = 25. Not chain folding, but the latter mechanism is supposed to be responsible for a crossover from short n-alkane ('wax') behaviour to high-n ('polymer') behaviour. In the latter case, the polymer only partly crystallizes. The formation of layered structures is a natural way to deal with entanglements that cannot be removed in time to form a single layer. Obviously kinetic effects will play a role in the process of unmixing, during which the polymer can be stretched and incorporated into a growing crystal while other fragments near entanglements remain amorphous. In this way we arrive at the general model for the chain conformations, as given in Figure 4.20c.

4.4.1   Example 1: Polyethylene (PE)

The x-ray signature of PE crystals has already been given in the WAXS results of Figure 4.18. Note the broad diffuse peak from the amorphous part of the sample with sharp peaks on top from the crystalline parts. Figure 4.21 gives the complementary SAXS data from the stacks of crystalline lamellae (Figure 4.20c). This peak is not sharp due to a relatively broad distribution of the corresponding lamellar periods.

The sharp wide-angle peaks of PE indicate an orthorhombic crystal structure, shown in Figure 4.22. The three-dimensional lattice is defined by a strict periodicity along the polymer chain in the *c*-direction, that depends on the chain conformation. In the present case the chain forms a planar zig-zag of carbon atoms, which is quite common. However, as we shall see in the next example, it can also be a helix, where the type of helicity depends on the chemical constitution of the chain. The chain packing is often not very specific, which results in a (pseudo) hexagonal pattern similar, as expected, to cylindrical rods. In polyamides (nylons), intrinsic interchain forces (hydrogen bonding) come into play, forcing the planar zig-zag chains into sheets.

Interestingly, PE can also form a stable hexagonal phase, either as a transient during crystallization or in equilibrium above 3.2 kbar (Rastogi et al., 1991). Upon cooling from the melt at high pressures, the hexagonal phase is obtained first. To transfer into the orthorhombic phase upon further lowering the temperature, the spacing of the (100) planes must decrease from

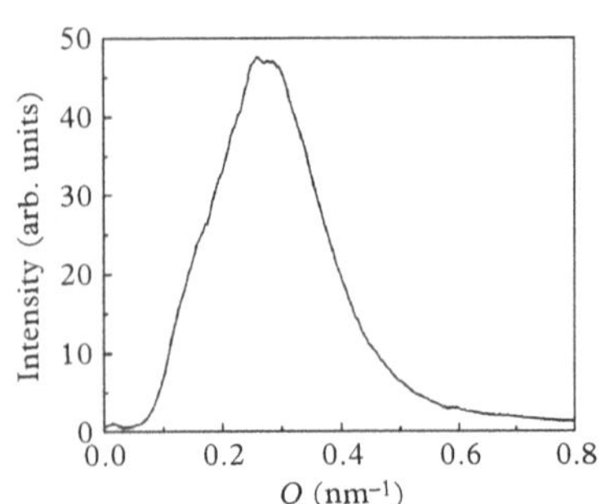

Figure 4.21 *SAXS of PE indicating the approximate periodicity of the stacked crystalline lamellae ('long spacing' of about 23 nm).*

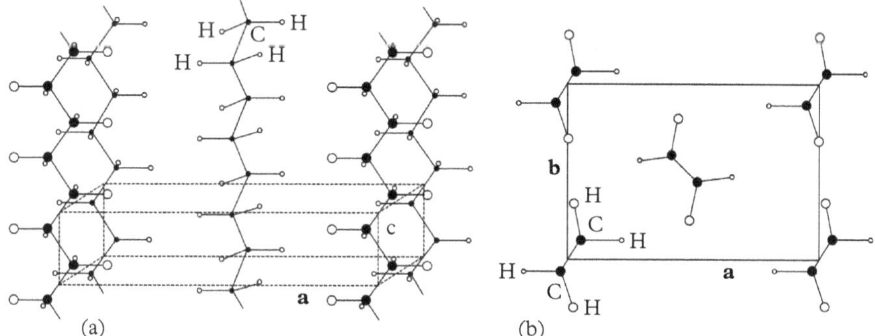

Figure 4.22 *Orthorombic crystal structure of PE (after Bunn, 1939) with a = 0.74 nm, b = 0.493 nm, and c = 0.254 nm. (a) General overview. (b) Projection along the c-axis.*

$a/b = \sqrt{3} = 1.73$ to the value 1.69, only a small change of a. In addition, the time-averaged equivalence of the hexagonal chains will be lost. The presence of a transient hexagonal phase supports the idea of crystallization of the fractional molecular bundles mentioned above. The transient phase has been put into use to control the uniformity of the final PE sample (Rastogi, Kurelec, and Lemstra, 1998).

4.4.2   Example 2: Isotactic polypropylene (iPP)

Next to PE, iPP is the second polymer of extraordinary commercial importance. The chain can have a helical structure, as shown in Figure 4.23. As a result, the crystal structure is somewhat more complicated than for PE: the standard α-form is monoclinic with $a = 0.665$ nm, $b = 2.10$ nm, $c = 6.50$ nm, and $\beta = 99.3°$ (Natta and Corradini, 1960). The melting point is about 170°C. In addition to the α-form, other crystal structures exist (β and γ; see Turner-Jones, Aizlewood, and Beckett, 1964). Numerous studies of iPP have been performed and we shall use iPP as a model example to illustrate some further aspects of polymer crystallization.

Obviously, the final properties of a polymeric material will depend strongly on the crystallization situation (type of crystals, crystalline vs amorphous fraction). In a typical crystallization experiment, the material is quenched from the isotropic melt to below the melting point. The speed and form of crystallization depend on the degree of supercooling. In addition, crystallization is strongly promoted by either the presence of nucleation agents[18] or by shear or flow.[19] These aspects have been widely studied because of their practical importance in the context of extrusion. In Figure 4.24 we illustrate the latter effect by measurements of the time dependence of the crystallization of iPP.

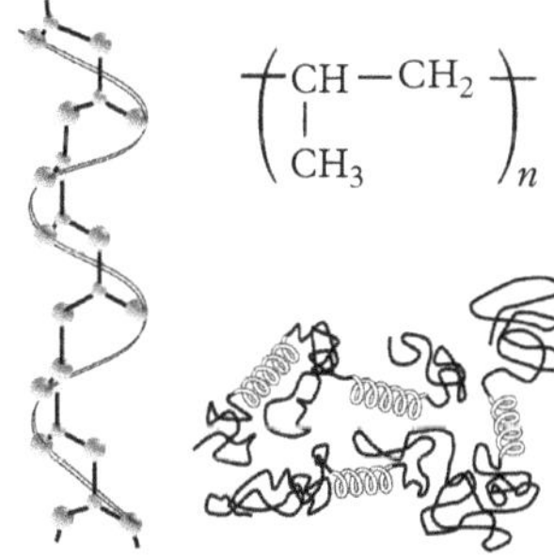

Figure 4.23 *Helical structure in the solid state of iPP (left) which returns short-range in the amorphous/liquid state (right).*

[18] See, for example, Byelov et al. (2008).

[19] See, for example, Kornfield, Kumaraswamy, and Issaian (2002).

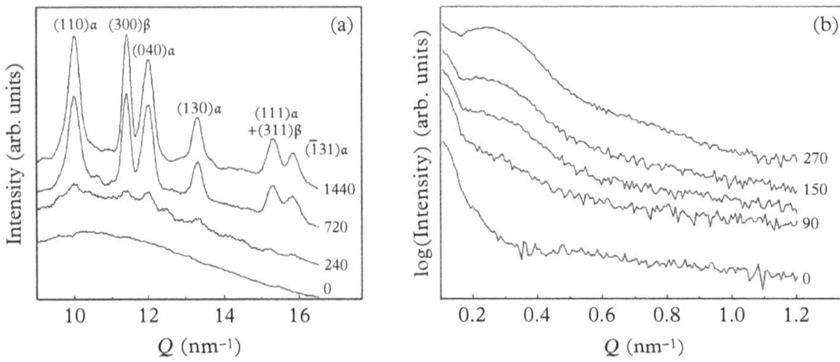

Figure 4.24 *Crystallization of sheared iPP as a function of time (indicated in seconds). The material was equilibrated at 210° C. After quick cooling to 140° C, a single rotation step shear was applied (1 s^{-1} and strain 1500%) and subsequently WAXS and SAXS were monitored simultaneously.*
(a) WAXS displaying mainly the monoclinic crystalline α-form.
(b) SAXS showing the 'long spacing' associated with alternating crystalline and amorphous lamellae. Note the logarithmic intensity scale.

4.4.3   Degree of crystallization

The degree of crystallinity of a polymer is an important quantity with respect to its final properties. It can be obtained from WAXS measurements by comparing the scattered intensity of the crystalline and amorphous part of the sample. The total scattered intensity in three-dimensional space can be written as

$$\int I(Q)dQ = \int \rho^2(r)d\mathbf{r}.$$

As the mean electron density of amorphous and crystalline parts are roughly the same, this can be written as

$$\int (I_{am} + I_{cr})dQ = \bar{\rho}^2 \int (dV_{am} + dV_{cr}) \propto V_{am} + V_{cr},$$

where V_{am} and V_{cr} are the appropriate volumes and $\bar{\rho}$ is the average electron density. We assume that the electron density distribution has spherical symmetry (disoriented 'powder' pattern). Then the total scattered intensity is obtained by integrating the intensity $I(Q)$ over the total reciprocal volume

$$\int I_{am/cr}(Q)4\pi Q^2 dQ \propto V_{am/cr},$$

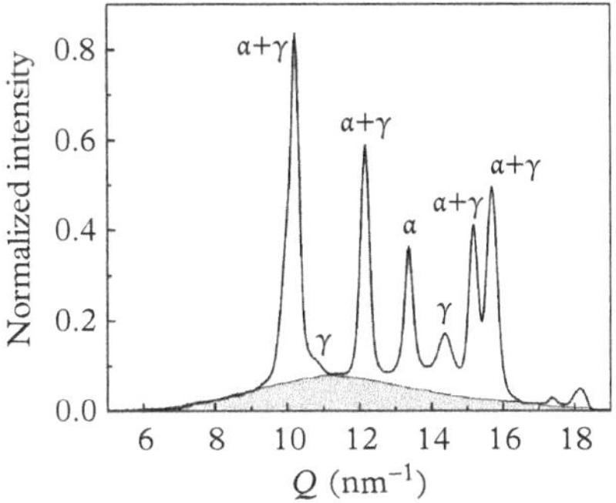

Figure 4.25 *Subtraction of the amorphous part from the x-ray pattern of crystallized iPP (after Van der Burgt et al., 2002). The present sample shows the α and γ crystal forms (no β form). The α peaks can be indexed by comparing with Figure 4.24a.*

where the index refers to either 'amorphous' or 'crystalline'. Hence the problem reduces to separating, in the total scattered intensity, the two different parts. This can be done using the fact that the crystalline scattering is concentrated in sharp diffraction peaks and that the amorphous scattering in a broad halo can be subtracted from the total scattering. Then the desired degree of crystallinity, X_{cr}, is given by

$$
X_{\mathrm{cr}} = \frac{\displaystyle\int I_{\mathrm{cr}}(Q)Q^2\,dQ}{\displaystyle\int I_{\mathrm{total}}(Q)Q^2\,dQ}, \tag{4.12}
$$

where the integration limits are over the total accessible Q-range. The crucial point is to obtain a smooth curve for the amorphous scattering (see Figure 4.25 for an example). Usually, the amorphous part is overestimated and thus X_{cr} is somewhat too low, because weak crystal reflections (especially at large and small angles) and crystal imperfections may be included in the amorphous part. Part of these problems can be avoided by measuring the purely amorphous compound, for example directly after quenching when still no indication of crystallinity is detected in the scattering. The problem of subtracting the amorphous signal is then reduced to the correct scaling of the amorphous signal.

4.4.4   The onset of crystallization

In the picture so far, crystallization starts with nucleation and growth of spherulites. For a further discussion of these processes we refer to the literature (see, for example, Strobl, 2007). However, a specific controversy started in the 1990s regarding the onset of crystallization. In the conventional model, nucleation is possible as soon as the temperature is decreased below the melting point and starts after an induction period. It is an activated process that requires sufficient

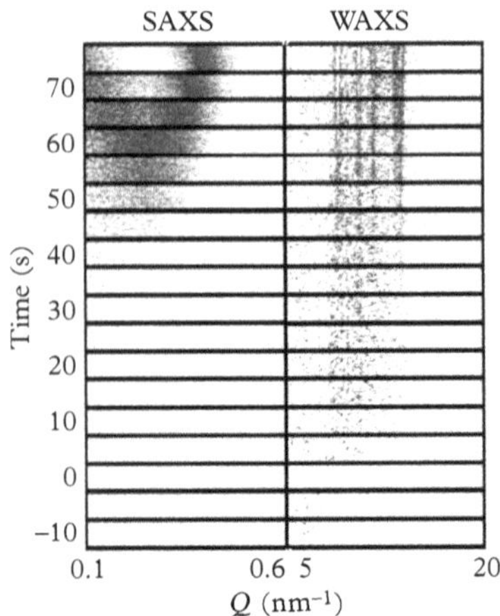

Figure 4.26 *The onset of crystalliza-tion in iPP (adapted from Panine et al., 2008, with permission from Elsevier.)*

supercooling before crystallization can take place in a reasonable timespan. During the following radial growth, for lamellar polymer crystals, the lateral at-tachment of chains is a determining process. In terms of x-ray characterization this means that first small crystallites are formed (to be observed by WAXS) and subsequently the first stacks of crystalline layers develop (leading to a SAXS sig-nal). The debate was triggered by reports that in iPP SAXS intensity appeared first, suggesting the existence of precursor structures (onset of some long-range order) in the melt before the development of crystals. This effect was attrib-uted to a coupling between density and chain conformation leading to spinodal decomposition—the spontaneous growth of fluctuations.

The above considerations generated a strong debate, focusing on the experi-mental evidence. Obviously, a decisive factor is the sensitivity of the (different) x-ray detectors used for WAXS and SAXS, respectively. Ultimate experiments were finally carried out at the ESRF in which, for both signals, almost single-photon sensitivity could be reached. The results are shown in Figure 4.26 and confirm the conventional picture (WAXS signal from nucleation earlier than SAXS). Though this resolves the specific point mentioned, it should be real-ized that in the broad context of polymer crystallization the possible role of precursors—like transient (meso)phases (Strobl 2009)—is still a topic.

Applications to Soft Matter

5

This chapter brings a wide selection of examples from soft matter, in which the knowledge of x-ray scattering of the previous chapters is applied. The choice of topics reflects, to some extent, personal interests and could have been different. More importantly, each of the sections treats a field that has been covered by full monographs. Hence, the basic information supplied for each topic is necessarily limited, and restricted to aspects crucial to the context of x-ray scattering. The text is not polluted with (too) many citations and the reader is referred to the general references for more details on the various soft matter systems. The central part of the chapter is preceded by a discussion of some practical aspects of x-ray experiments. The chapter finishes with a case study on the role of order and frustration when different length scales are combined in a single soft matter system.

5.1 Instrumentation, SAXS, and WAXS

The principle of a simple x-ray setup was given in Figure 2.4. The x-ray beam hits the sample and the diffracted beam is measured by a detector (point, linear, or two-dimensional) at some distance from the transmitted beam. Any x-ray peak on the detector is characterized by its distance X from the centre of the detector. Knowing the sample–detector distance, L, we can calculate the scattering angle from $\tan 2\theta = X/L$. From the definition of Q or from the Bragg equation, the corresponding periodicity, d, can be calculated. So far, we simply assumed a nicely parallel incident beam, but reality is not so simple. The process of generating x-rays as described in Chapter 1 will give, in general, a divergent beam[1] instead of the required highly collimated parallel beam. In the absence of ready-to-use x-ray optics, collimation is complicated and conventionally realized by a slit system. A typical three-slit configuration as standard in use for small scattering angles, is shown in Figure 5.1.

During the last decades, multilayer systems with a parabolic curvature (Göbel mirrors) have come into use. These mirrors are highly sophisticated devices that have a continuous variation of the Bragg angle over its parabolic surface. Hence, Bragg reflections can be obtained over a certain angular range, converting

[1] Note that this does not apply to synchrotron radiation (see the end of Section 1.2).

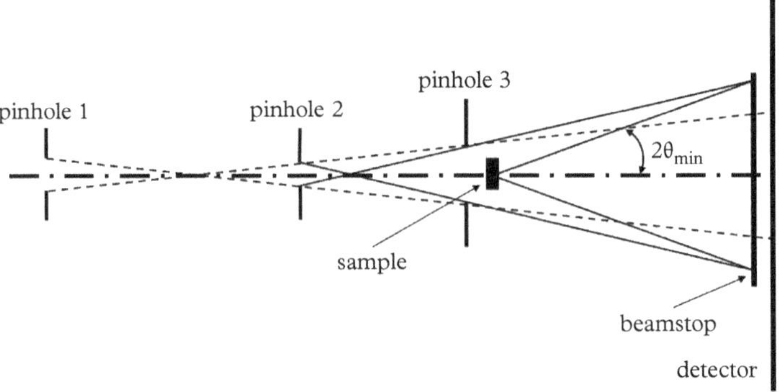

Figure 5.1 *Pin-hole system for collimating an x-ray beam.*
The beam is defined by pinholes 1 and 2. Pinhole 3 removes any scattering from the edges of the previous ones.

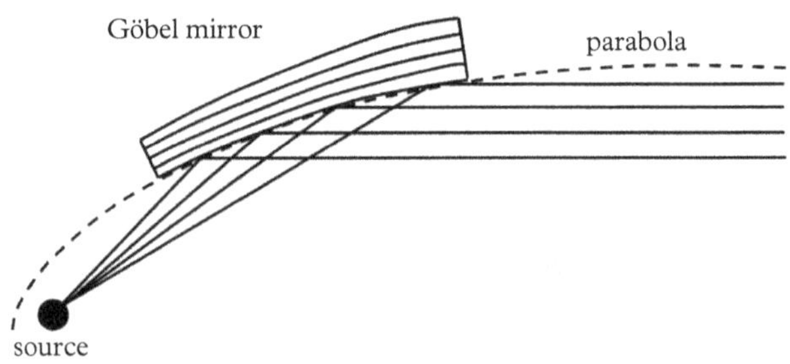

Figure 5.2 *With a small micro-source in the focal point of the parabolic surface of a Göbel mirror, Bragg reflections from the graded multilayer system convert part of a divergent beam into a parallel one.*

a divergent beam into an (almost) parallel one (see Figure 5.2). The gain in intensity over slit collimation is about an order of magnitude with a remaining divergence of typically <50 mdeg. The combination of two perpendicular mirrors and an intense micro-source has led to a revival of the conventional type of setup of Figure 5.1.[2]

In the transmission geometry as discussed, a sample holder may consist of a metal washer with an opening slightly larger than the beam size. Two thin kapton sheets glued to the washer serve to confine the sample. Alternatively, thin-walled capillary tubes from quartz may be used. Obviously for liquid sample these capillaries are the only option.

In practice, small-angle (SAXS) and wide-angle (WAXS) scattering are distinguished. Since the measured scattering is a description of the sample structure in reciprocal space, small angles correspond to large periodicities and large angles to small ones. For SAXS usually the midpoint of a two-dimensional detector is positioned at the centre of the incoming beam, the detector being protected from the high intensity by a beamstop (Figure 5.1). To use this setup (Figure 2.4 and Figure 5.1) for WAXS, it is sufficient to reduce the sample–detector distance to reach the right Q-range. In this way, a two-dimensional picture can be obtained, like the one shown in Figure 4.18 for PE. The nice thing is, of course, that with a two-dimensional detector anisotropies like alignment can also be determined.

<hr>

[2] Using an elliptic mirror with the point source in the first focal point, the resulting output will be concentrated in the second focal point of the ellipse.

For WAXS of multicrystalline or powder systems, often an x-ray source is used that is shaped as a thin line instead of a point source.[3] For these applications parallelism of the incoming beam is not crucial, because all scattering angles are present in the sample anyhow, and a Göbel mirror is not needed. Common is the so-called Bragg–Bretano geometry in which a mildly divergent beam is reflected from the sample and subsequently refocused on a one-dimensional detector. This process is illustrated in Figure 5.3, in which source, sample, and detector are on the same focusing circle. The x-rays 'reflected' from the sample are directed to their corresponding positions on the circle.

The experimental division between SAXS and WAXS in terms of small and wide scattering angles is not always useful from a conceptual point of view. A more meaningful division is between measurements of non-interacting particles (form factor, as discussed in Chapter 2) and of the structure factor (diffraction, as discussed in Chapter 4). The first situation usually corresponds to small angles (SAXS). The second situation involves structure, for conventional crystalline systems at periods of the order of 0.1–1 nm, which is indeed at wide angles (say $5° < 2\theta < 70°$). However, many soft-matter systems show structural features on much larger length scales leading to x-ray intensity at smaller angles outside the conventional WAXS range. Examples comprise liquid crystals that are ordered on a scale of several nm (intermediate angles between classical WAXS and SAXS), and block copolymers that phase-separate into structures of the order of 1–20 nm (SAXS). We shall discuss several examples of this type in the following paragraphs. Nevertheless, in the literature the latter structure measurements are usually indicated as SAXS results.

The use of standard experimental setups for SAXS (like Figure 5.1) and for WAXS (like Figure 5.3) creates problems for measuring the intermediate range. For example, liquid-crystal order can be observed at the large-angle side of a SAXS setup (which may require inconvenient shortening of the sample–detector distance) or at the small-angle side of a standard WAXS setup (measuring inconveniently close to the direct beam). The most suitable setups available these days allow for a continuously variable sample–detector distance and thus cover the full range from WAXS via intermediate angles to (ultra)SAXS.

5.2 Examples

5.2.1 Liquid crystal phases

As noted before, the anisotropy and low-dimensional aspects of liquid crystals provide some excellent models to illustrate various aspects of ordering and their x-ray signature. In this section we shall treat successively discotic (disc-like) liquid crystal systems, ordering inside the smectic layers, and finally the nematic–smectic phase transition.

Just as rod-like molecules can orient their long molecular axes to form a nematic phase, disc-like molecules can do the same with respect to the normal to

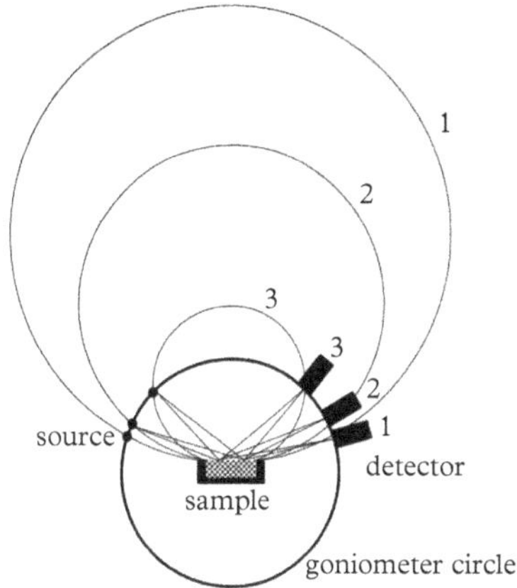

Figure 5.3 *Multicrystalline diffraction in the reflective mode (Bragg–Bretano geometry) using x-rays from a line source perpendicular to the plane of drawing. In this geometry, the beam incident on the sample is refocused at the linear detector.*

[3] A typical dimension of the focal spot of a commercial line source is $12 \times 0.04\,\mathrm{mm}^2$.

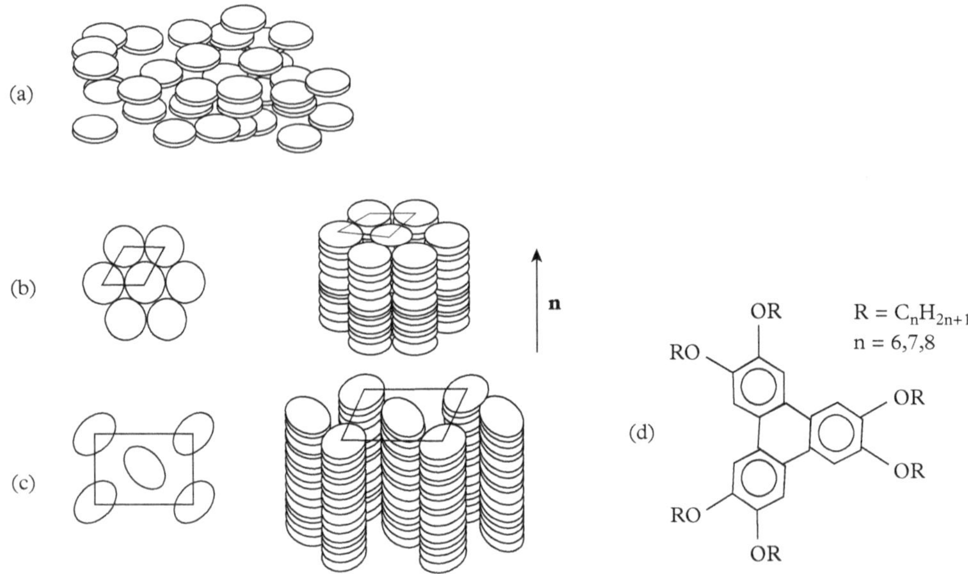

Figure 5.4 *Plate-like molecules in (a) a discotic nematic phase, (b) a hexagonal columnar phase, and (c) a rectangular columnar phase. (d) Classical example of a discotic compound.*

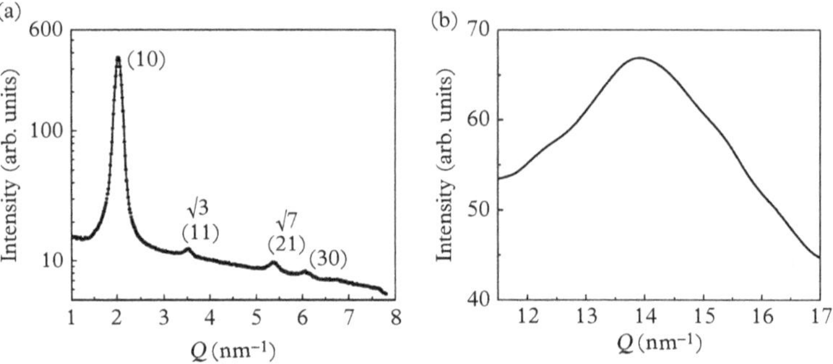

Figure 5.5 *X-ray patterns of a hexagonal discotic phase with disordered columns. (a) SAXS. (b) WAXS. For these two-dimensional structures we assign (hk) to each peak and disregard the (irrelevant) third Miller index.*

the discs. They form discotic liquid crystals, the simplest one being, again, a nematic phase (see Figure 5.4a). It is similar for rod-like molecules, but the preferred uniaxial direction (director **n**) is now along the average normal to the plates. As a result, for a uniformly oriented sample, the birefringence—that is positive for rod-like systems—is negative. In addition, the discs can order in columns that align with their axes and may be disordered along the column axis (Figure 5.4b, c). Fig 5.4d gives an example of a classical discotic molecule, but many more (exotic) examples can be found.

In the simplest case of a columnar phase (Figure 5.4b, diffractogram in Figure 5.5) the intercolumnar order is hexagonal with peaks following the ratio $Q/Q^* = 1 : \sqrt{3} : \sqrt{7} : 3$ (see Tables 4.2, page 59, and 5.1, page 88) and a fundamental dimension $a = 2\pi/Q_{10} = 2\pi/2.0 = 3.1$ nm. From the broad peak at larger Q-values (Figure 5.5b) we conclude to disorder inside the columns ('liquid' columns), just as the simple smectic-A phase has 'liquid' layers. The average distance between discs is $2\pi/13.9 = 0.45$ nm.

Obviously, with the columns as constituting units in a two-dimensional packing, potentially a whole series of phases is possible, each corresponding to one of the two-dimensional lattices of Figure 4.3. As a second example, a centred rectangular lattice is shown in Figure 5.4c. In this particular picture the deviations from circular symmetry stem from a tilt of the discs in the columns, but alternatively the molecules themselves can deviate from circular symmetry in the form of more rectangular plates. The x-ray signature is characterized by a splitting of the main fundamental peak, as shown in Figure 5.6.

Finally, we mention the possibility of a transition from disordered columns to ordered ones in a hexagonal columnar phase (Fontes, Heiney, and de Jeu, 1988). This poses the question as to whether neighbouring columns lock-in, leading to three-dimensional order (a crystal). Otherwise the columns are still free to slide along each other, leading to a two-dimensional structure of ordered one-dimensional cylinders that do not combine to become a three-dimensional solid. It seems that this is often the case. Whether solidification can happen to reach the final crystalline state is often not clear for the resulting strongly waxy systems.

Let us return to smectic phases of elongated molecules. Upon lowering the temperature crystalline order can develop inside the (liquid) smectic layers. To demonstrate these effects, consider the in-plane x-ray diffraction peaks of Figure 5.7 (compare with Figure 3.11). In (a) we start with the standard smectic-A phase: 'liquid' layers, giving a broad peak. Below the diffractogram, a simple model is pictured for the local structure. The projection of a liquid-crystal molecule along its long axis is taken as anisotropic because of the rather flat aromatic part. However, any structure in the smectic layer will only be short-range: disordered layers. In (b) additional positional order is present. Moreover, in-plane a local herringbone orientation has developed that is, however, not fixed yet over

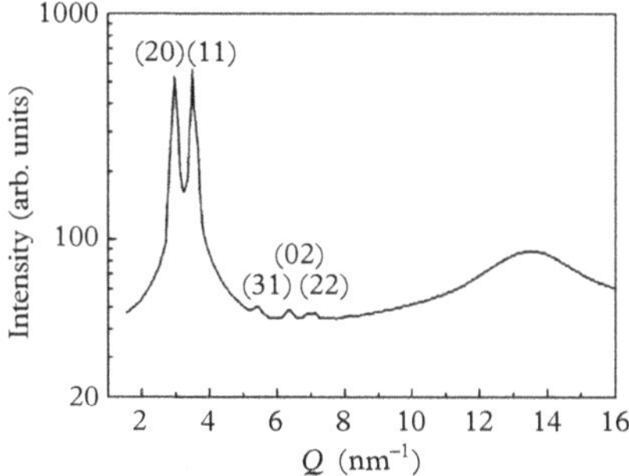

Figure 5.6 *X-ray pattern of the centred rectangular columnar phase of rufigallol hexa-n-octanoate (Billard et al., 1981; Chandrasekhar et al., 2002). Note that the large Q-range covers both the SAXS and the WAXS regions.*

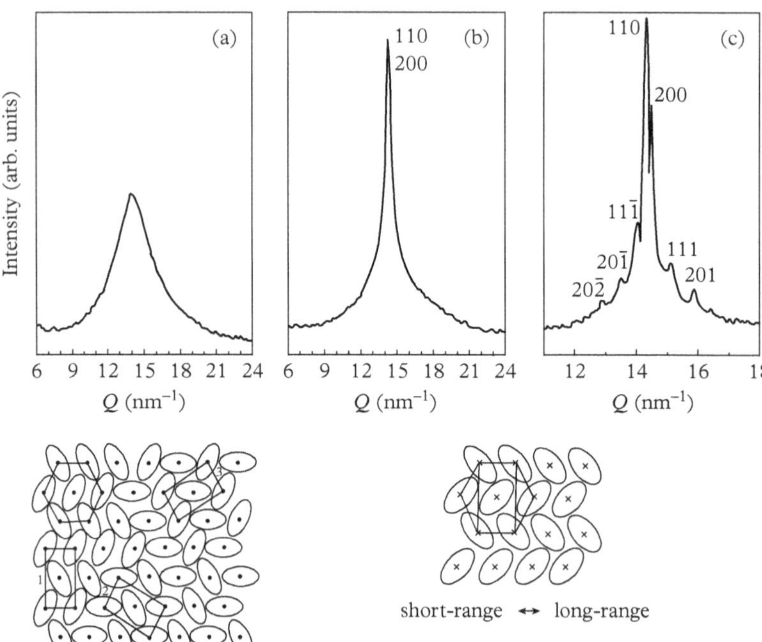

Figure 5.7 *Top: In-plane order of various types of smectic phase: (a) liquid (smectic-A), (b) hexagonal (crystal-B), (c) herringbone–orthorhombic (crystal-E) (after de Jeu and de Poorter, 1977). Bottom: Representation of the local in-plane molecular order for cases (a) and (b, c). The herringbone orientations are short-range for (b) and become long-range for (c).*

longer distances. As a result, the local symmetry is hexagonal, giving a single sharp peak (smectic-B). Finally, in (c), the herringbone orientations also becomes long-range, leading to the biaxial smectic-E phase.

Again, the situation is more complicated then it seems at first sight. For true three-dimensional crystallinity, the intra- and inter-layer order must be coupled. Otherwise one could imagine quasi-crystalline layers that still translate and rotate with respect to each other. In the latter case, x-ray peaks indexed as *hkl* separate into combinations *hk*0 (inside the layers) and 00*l* (from the layering itself). In that situation, a two-dimensional structure plus a one-dimensional one does not necessarily give order in three dimensions![4] In fact for most cases reported in older literature to be smectic-B there is a (weak) coupling and x-ray intensity is observed for non-zero values in both *hk* and *l* (not seen in Figure 5.7b) (Leadbetter, Frost, and Mazid, 1979). Therefore, the indication smectic-B is not used anymore, but crystalline-B (Cr-B) is.[5] For the herringbone structure of Figure 5.7c, the three-dimensional coupling is clear from indices like (111) and (201): Cr-E phase. These Cr-B and Cr-E phases have been called lamellar plastic crystals.[6]

To finish this section we discuss the phase transition nematic to smectic-A. As mentioned in Chapter 3, because this transition corresponds to melting/crystallization in one dimension, it is of great fundamental interest. As a result is has been studied extensively, both theoretically and experimentally. Interestingly, the phase transition varies in various compounds from first-order to second-order. In the latter case, the smectic order parameter changes continuously at the

---

[4] Compare with the earlier discussion of sliding columns.

[5] Smectic-B (SmB) is reserved for the hexatic smectic-B phase, discussed in Section 3.4.2.

[6] We mention in passing tilted smectic phases, in which the local director makes an angle with the layer normal. In combination with the symmetry elements discussed already, this leads to a 'zoo' of smectic phases that we do not want to include here.

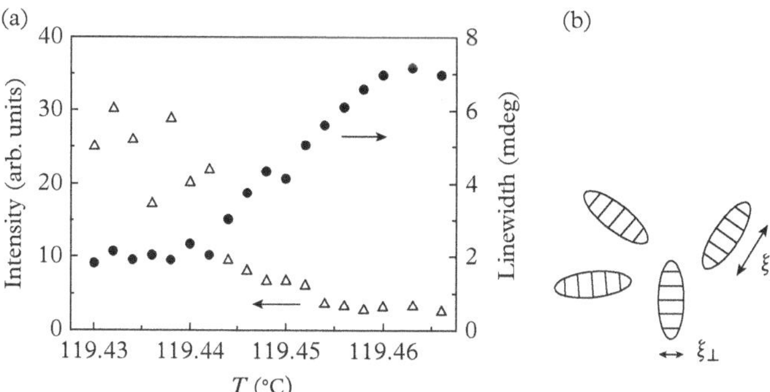

Figure 5.8 *(a) High-resolution x-ray results around the nematic–smectic-A phase transition at 119.443 °C of the compound of Figure 6.8b. (b) Model of pre-transitional smectic 'droplets' in the nematic phase. (Bouwman and de Jeu, 1994.)*

phase transition. This is demonstrated in Figure 5.8a. Upon heating across the smectic-nematic phase transition the intensity of the smectic peak goes down, while the line width (FWHM) increases, as expected for the liquid-like peak in the nematic phase. Within the limits of the (high) resolution of the measurement, there is no indication of any discontinuity.

A simple model of the phase transition is as follows. If we approach the transition from the nematic side (high temperatures), pre-transitional smectic order develops in the nematic phase. This can be imagined as smectic 'droplets' (see Figure 5.8b, but as a dynamic process, not static) that increase in size upon approaching the phase transition. The average size of these (anisotropic) smectic regions can be described by two correlation lengths that increase upon approaching the phase transition. This will be directly reflected in the FWHM of the x-ray line shape (Bouwman and de Jeu, 1993, and references therein).

5.2.2 Colloids

Colloids consist of assemblies of particles that are substantially larger than atomic scales but still much smaller than a macroscopic size. Typical colloidal dimensions are between 1 nm and 1 μm. They form an interesting class of materials that bridge the gap between the molecular and macroscopic world. We encountered an example of SAXS of a dilute solution of non-interacting (spherical) colloids in Section 2.3.3. Concentrated suspensions of colloids display transitions from fluid-like structures to ones exhibiting long-range positional order (colloidal crystals) and/or orientational order (colloidal liquid crystals). As their dimensions are significantly larger than conventional solvent molecules, in x-ray scattering the medium can be taken as a homogeneous background. Simple colloidal spheres can form macroscopic crystals of which the structure can be determined by optical methods (confocal microscopy) instead of x-rays (Yethhiraj and van Blaaderen, 2003). As model systems, colloids are useful in helping to extend our knowledge of (soft) condensed matter problems like freezing/melting, glass transitions, and the effects of external fields. However, a complicating factor is the dependence of their phase behaviour on the degree of dispersity.

If anisotropic colloidal particles like rods and platelets are dispersed in a liquid medium, they can self-organize into various liquid-crystalline structures (Kuijk, van Blaaderen, and Imhof, 2011). Mineral colloidal liquid crystals are especially interesting because they are electron-rich and therefore highly susceptible to external influences like electric and magnetic fields. Furthermore, they are thermally stable and liquid crystal phases can be observed over a wide range of temperatures (Davidson and Gabriel, 2005).

A well-known example of a mineral liquid crystal is goethite (α-FeOOH).[7] It gives in dispersion orthorhombic crystals in which the preferred direction of growth leads to board-like particles (see Figure 5.9) with a permanent magnetic moment along their long axis (Lemaire et al., 2002). These in turn are the constituting units (orders of magnitude larger than molecular systems) of liquid crystalline phases. In Figure 5.10 the x-ray patterns are displayed of various magnetically aligned liquid crystal phases of goethite, together with models of their organization. Interestingly, the overall nature of the smectic order is not different from that of smectic phases from molecular systems, discussed in Section 3.3.2. However, some questions remain regarding possible differences in elastic behaviour due to the large units of the goethite smectic phase compared with conventional smectic phases.

[7] It is named after Johann Wolfgang von Goethe, the German philosopher and poet who was also a mineralogist. The compound has been used as a pigment since prehistoric times.

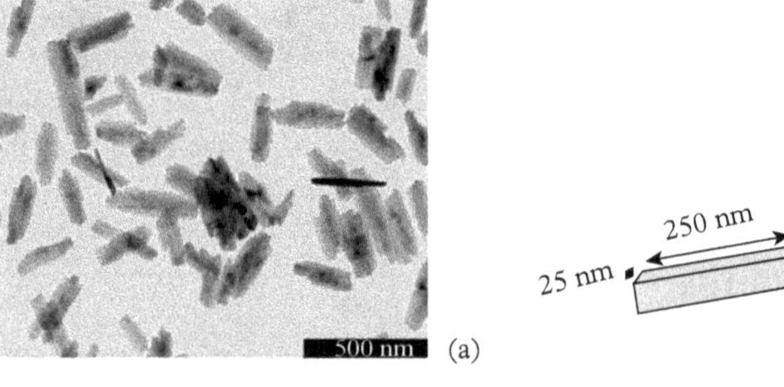

Figure 5.9 *Goethite particles. (a) TEM picture. (b) Typical dimensions. (Courtesy Van 't Hoff Laboratory, University of Utrecht, Netherlands.)*

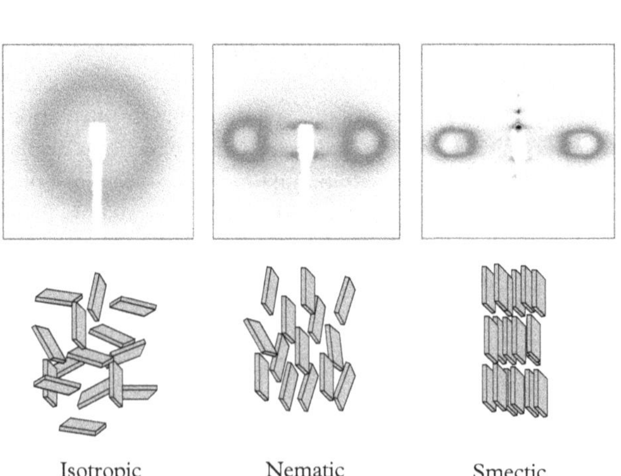

Figure 5.10 *X-ray patterns in the isotropic, nematic, and smectic phase of goethite particles oriented by a magnetic field (after Van den Pol et al., 2008).*

(a)

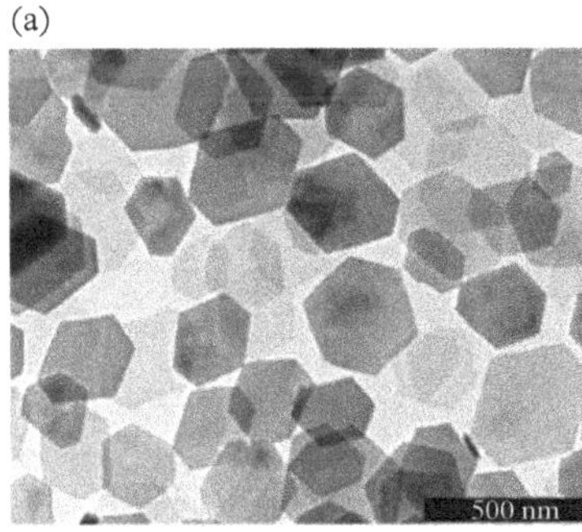

(b)

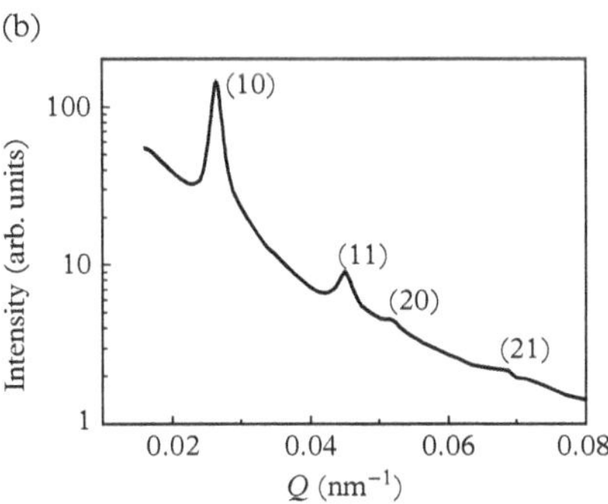

Figure 5.11 *Colloidal gibbsite platelets. (a) Transmission electron micrograph. (b) Diffractogram indicating a columnar hexagonal phase. (Adapted with permission from Mourad et al., 2009. Copyright American Chemical Society.)*

Not surprisingly after the previous example, platelet colloids can form a discotic phase. Figure 5.11a shows an electron micrograph of discs of the order of a few hundred nm diameter, and Figure 5.11b shows the corresponding x-ray picture in the small-angle region. The result is very similar to Figure 5.5a for molecular discs, but the Q-range is much smaller. For these colloid platelets the fundamental dimension is about hundred times larger than encountered earlier: $a = 2\pi/Q_{10} = 2\pi/0.028 = 0.22\ \mu$m.

5.2.3   Amphiphilic compounds

Amphiphilic compounds or surfactants consist of elongated molecules with a polar head and an aliphatic (non-polar) tail. They can be found in many cosmetic and pharmaceutical products and occur naturally in living systems. Their specific behaviour is directly related to the well-known framework of water and hydrocarbons (oil) that do not mix. If we try, the gain in entropy – that is at the origin of mixing – is more than offset by the unfavourable interactions between the two. In the case of amphiphiles the two states are unified in one molecule: the polar part prefers not to mix with the non-polar part, even though they are connected. In solution, the solvent can be a good or a bad one for either the polar or non-polar part. As a result, surfactants tend to form aggregates leading to rich phase diagrams in dependence of polarity, solvent, and concentration. The hydrophobic part of amphiphilic compounds is often a long straight hydrocarbon chain $CH_3(CH_2)_n$, with n > 4, while the hydrophilic part can be a polar functional group such as COOH or a small ion, for example, COO^- or $N(CH_3)_3^+$. Two well-known examples of surfactants are pictured in Figure 5.12. Soaps like SDS are among the first and best studied examples. Phospholipid surfactants are of great interest because they are the primary constituents of cell walls. Figure 5.13 gives a couple of examples of structures formed by surfactants in solution, illustrating, also, the additional role of the geometry of the molecules.

Figure 5.14 shows a generic phase diagram of a surfactant in water. At high and low water concentration one finds a (normal or inverse) micellar structure.

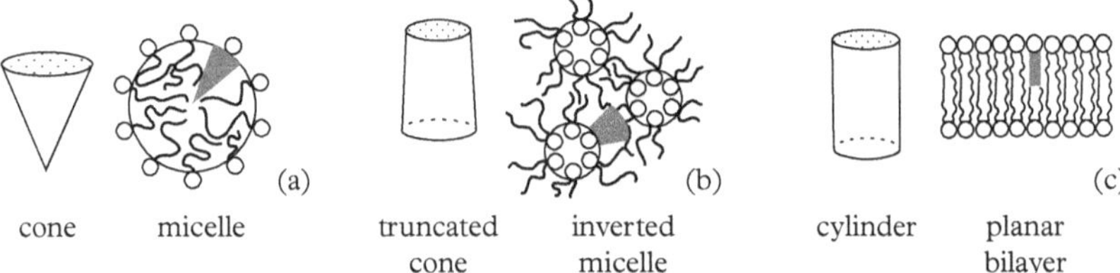

Figure 5.12 *Two examples of a surfactant: SDS, a soap with a single tail and DMPC, a phospholipid with two tails.*[8]

Figure 5.13 *Different local structures formed by lipids. (a) Micelle M. (b) Inverted micelle M_{inv}. (c) Flat bilayer L_α. The average molecular shape favours the particular structure.*

Above a critical micellar concentration (CMC) further addition of surfactants makes more micelles. The micelles interact with each other and have a tendency to form ordered phases (lyotropic liquid crystals) at higher volume fractions.[9] In that sense, they are similar to colloids. However, unlike colloids, micelles can change their shape as well as their size in reaction to their interactions. At around 50 percent water, a lamellar system is obtained in which water is sandwiched between bilayers of amphiphilic molecules. At higher or lower water concentration the spherical micelles (either regular or inverse) become elongated and form cylinders that can arrange in a hexagonal structure.

Concerning the phase diagram of Figure 5.14, a few remarks are appropriate.

- Obviously not all phases occur in all types of system.

- In reality there will be quite some temperature dependence and the phase boundaries may deviate strongly from vertical.

- The micelles can develop anisotropic shapes that may pack in various ways that differ for normal micelles (oil core in water) and inverse micelles (water core in oil).

- Figure 5.14 is only a cut through a more complex three-phase diagram with water, oil, and surfactant at the three corners.

- As a result of the previous point, realistic phase diagrams are usually more intricate and additional isotropic phases could intervene (a-d in Figure 5.14). These may be regular cubic (a and d) or bicontinuous cubic (b and c, see example below).

X-ray diffraction is an important tool to identify the various phases. Table 5.1 provides the information used to decide on a certain phase from the Q-vales of the observed peaks. In Figure 5.15 examples are displayed of x-ray

[8] SDS stands for sodium dodecyl sulphate and DMPC stands for 1,2-dimyristoyl-*sn*-glycero-3-phosphocholine.

[9] Some names in use for the different phases go back to nineteenth century soap boilers ('middle phase' for the H_1 phase, 'neat phase' for the L_α phase).

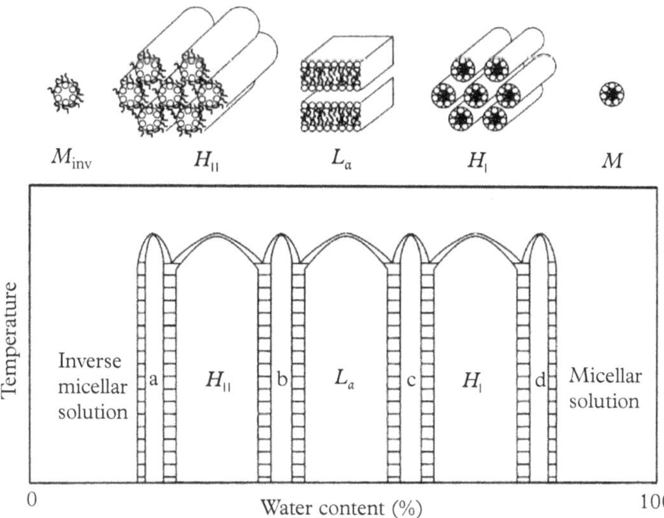

Figure 5.14 *Generic phase diagram of a surfactant in water with schematic representation of the phases (after Seddon, 1990). The intermediate phases indicated as a–d are isotropic (see text).*

patterns associated with amphiphiles for which some of the phases of the system monoglyceride–water have been chosen.

In particular, we note in (a) a simple lamellar phase and in (d) a hexagonal phase characterized by a peak at $\sqrt{3}$. In addition, in (b) and (c), two bicontinuous cubic phases are shown. In that case, the bilayer network is distorted and divides three-dimensional space into two interconnected domains. In fact, several of these types of phase exist with different internal symmetry. The two relevant ones from Figure 5.15 are illustrated in Figure 5.16. We shall also encounter such phases in Section 5.2.5 about block copolymers. Shear rheology shows that each of the phases has a specific rheological signature, characteristic of the topology

Table 5.1 *Peak positions relative to the fundamental one at Q^*, as derived from Table 4.2.*

Structure	Ratio Q/Q^*
Lamellar	$1 : 2 : 3 : 4 : 5 : 6$
Hexagonal	$1 : \sqrt{3} : 2 = \sqrt{4} : \sqrt{7} : 3 = \sqrt{9} : 2\sqrt{3} = \sqrt{12} : \sqrt{13} : 4 = \sqrt{16}$
Cubic bcc	$1 : \sqrt{2} : \sqrt{3} : 2 = \sqrt{4} : \sqrt{5} : \sqrt{6} : \sqrt{7} : 2\sqrt{2} = \sqrt{8} : 3 = \sqrt{9}$
Cubic fcc	$\sqrt{3} : 2 = \sqrt{4} : \sqrt{8} : \sqrt{11} : 2\sqrt{3} = \sqrt{12} : 4 = \sqrt{16} : \sqrt{19}$
Cubic gyroid	$\sqrt{3} : 2 = \sqrt{4} : \sqrt{7} : \sqrt{8} : \sqrt{10} : \sqrt{11} : 2\sqrt{3} = \sqrt{12} : \sqrt{13} : \sqrt{15}$
Cubic double diamond	$\sqrt{2} : \sqrt{3} : 2 = \sqrt{4} : \sqrt{6} : \sqrt{8} : 3 = \sqrt{9} : \sqrt{10} : \sqrt{11} : 2\sqrt{3} = \sqrt{12}$

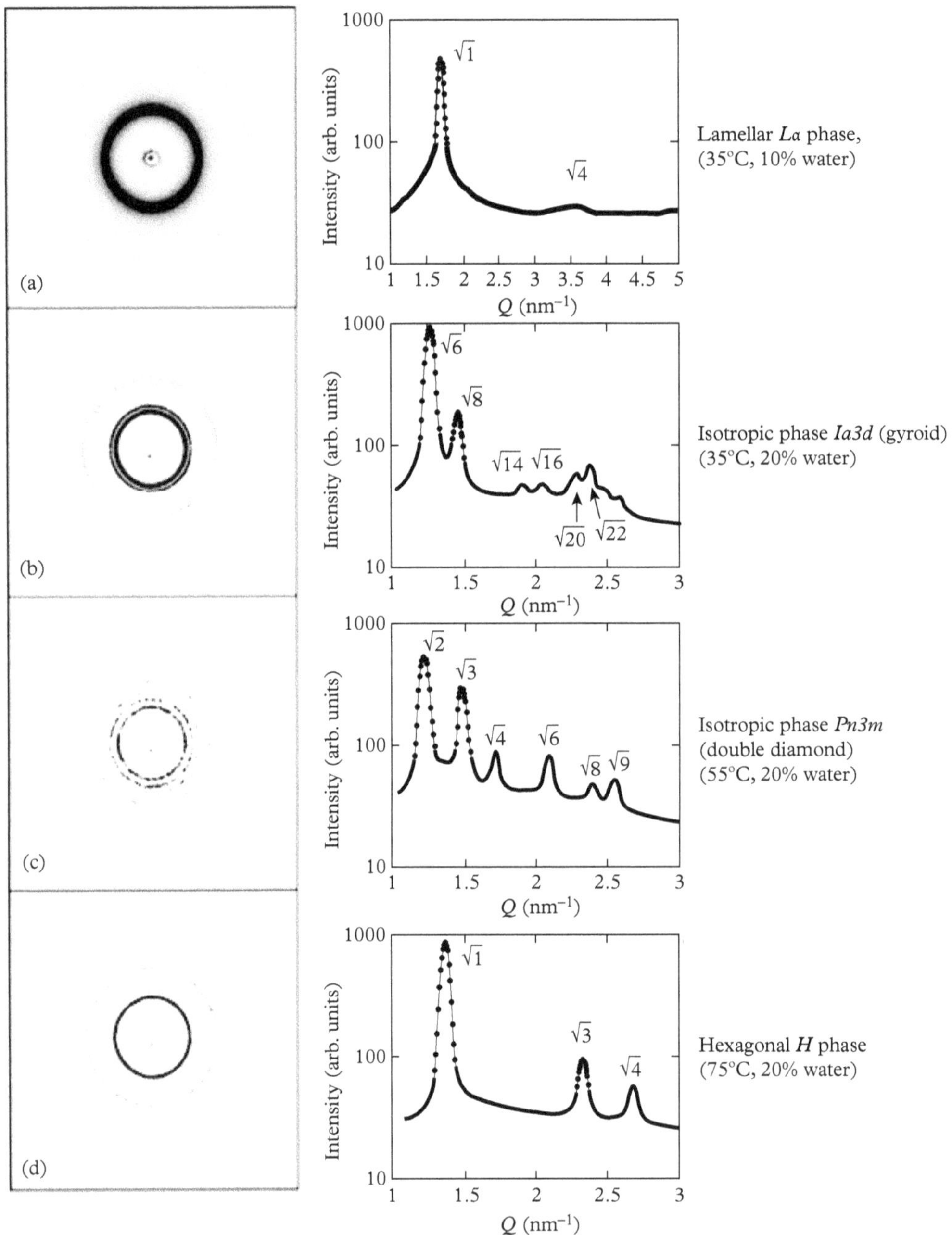

Figure 5.15 *SAXS structures for the various liquid crystal phases of the system monoglyceride–water. (Reprinted with permission from Mezzenga et al., 2005. Copyright American Chemical Society.)*

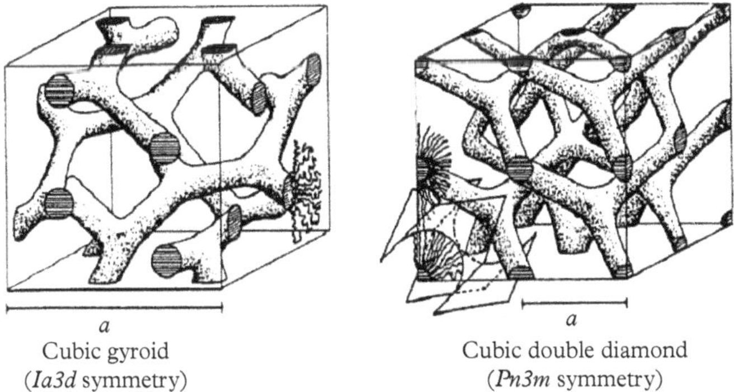

a	*a*
Cubic gyroid	Cubic double diamond
(*Ia3d* symmetry)	(*Pn3m* symmetry)

Figure 5.16 *Models of the bicontinuous cubic phases of Figure 5.15. (Adapted from Seddon et al., 2000, with permission of the PCCP Owner Societies.)*

of the structure. This also allows us to characterize the order–order transitions lamellar-to-cubic and cubic-to-hexagonal, and those between the bicontinuous cubic phases.

5.2.4 Biomembranes

A natural extension of Section 5.2.3 is to move to bilayer systems (membranes) that consist mainly of phospholipids. The basic principle is once more illustrated in Figure 5.17, where two simple membranes are pictured that can interact via a water environment. Multilayer membranes consist of equally spaced fluid bilayer lipid systems embedded in water.

A biomembrane is a more complex multicomponent system that contains specialized proteins that may mediate specific interactions and structural changes in the membrane. Obviously, that is beyond the scope of this short section. The lipid bilayer and interbilayer interactions are basic to all biomembranes. The forces that come into play comprise long-range Van der Waals interactions, short-range hydratic repulsion, and screened electrostatic forces. Moreover, thermally induced out-of-plane layer fluctuations lead to repulsive interactions because of steric hindrance in multilayer systems (Helfrich-effect). In the following, we shall discuss one rather specific case of DNA electrostatically adsorbed at cationic lipid surfaces. For a broader overview we refer the reader to Tresset (2009).

When DNA is complexed with cationic membranes, spontaneously a multilayer assembly is formed of DNA and bilayer membranes. These DNA–cationic membrane complexes have attracted attention because they can mimic certain characteristics of natural viruses by being efficient chemical carriers of genes (DNA sections) for delivery in cells (Safinya, 2001). This allows us to study fundamental issues of the structures and the intermolecular interactions in DNA-membrane assemblies that are of fundamental biological interest. Figure 5.18a visualizes such a DNA complex in a 50 percent binary mixture of the neutral and cationic lipids, dioleoyl-phosphatidylcholine (DOPC), and

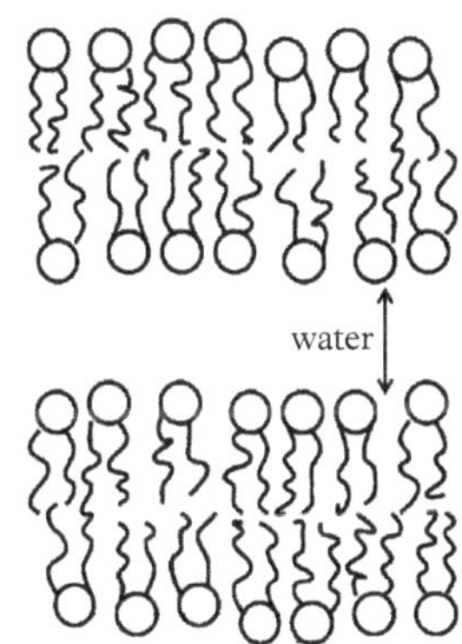

Figure 5.17 *Cartoon of two interacting fluctuating membranes.*

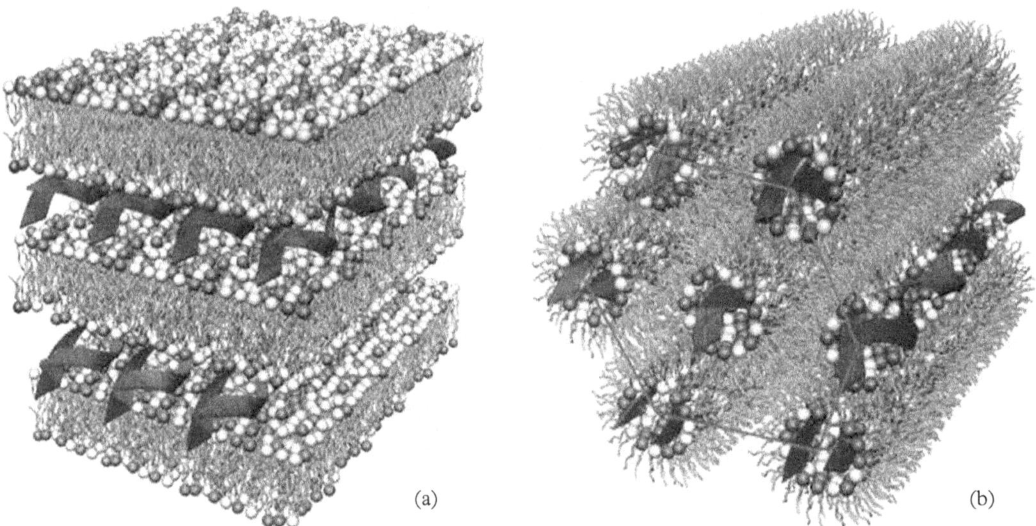

Figure 5.18 *Cationic lipid–DNA complexes. (a) Local arrangement in a lamellar phase with alternating lipid bilayer and DNA monolayer (Rädler et al., 1997). (b) Similar complexes in an inverted hexagonal phase. The cylinders consist of DNA coated with a lipid monolayer. (Reprinted with permission from Koltover et al., 1998. Copyright American Association for the Advancement of Science.)*

dioleoyl-trimethylammonium propane (DOTAP), respectively. In Figure 5.18b, a similar situation is depicted for a hexagonal structure. These are the structures we shall discuss hereafter from an x-ray point of view.

Figure 5.19a displays the SAXS results for the lamellar situation that can be understood from the schematic representation in Figure 5.19b. The parallel DNA chains form a periodic one-dimensional lattice between the stacked two-dimensional lipid sheets. The local structure of the self-assembled DNA-lipid complexes is characterized by a defined membrane repeat distance,

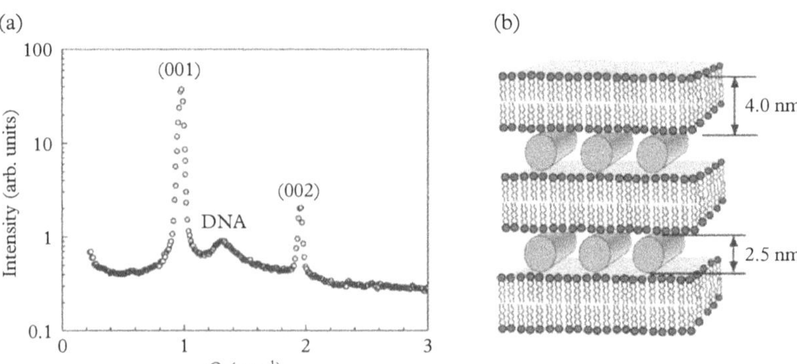

Figure 5.19 *(a) SAXS pattern of the membrane complex of Figure 5.18a, that can be understood from the simplified model in (b).*

$d_{\mathrm{mem}} = 2\pi/Q_{001} = 6.5\,\mathrm{nm}$ for the present DOTAP–DOPC ratio of about 50 percent. The total period consists of a lipid bilayer (around 4 nm, which can be measured by x-rays in the absence of DNA) and a DNA monolayer (about 2.5 nm, corresponding to B-form DNA with a hydration layer). In the absence of DNA, the multi-lamellar L_α phase of the charged lipid–water mixture would give rise to very large interlayer spacings of order 100 nm (at 99 percent water), due to long-range electrostatic interactions. The DNA that is condensed on the cationic lipid layer strongly screens this interaction, leading to a collapse of bilayers into condensed multi-layers. The broad peak in the middle arises from the two-dimensional in-plane packing of the DNA strands and gives an average DNA interchain spacing $d_{\mathrm{DNA}} = 4.8\,\mathrm{nm}$. The DNA peaks exhibit considerable line broadening, indicative of disorder in the columnar DNA lattice. No positional or orientational correlation exists between the DNA chains across adjacent lipid layers ('one-dimensional smectic').

Addition of the cone-shaped helper-lipid DOPE[10] causes curvature of the cationic lipid monolayers resulting in a transition to the hexagonal phase of Figure 5.18b. The x-ray signature of this transition is illustrated in Figure 5.20. The lower figure is essentially the same as Figure 5.19, the upper one can be well described on the basis of a two-dimensional hexagonal lattice with a unit cell spacing of $a = (4\pi/\sqrt{3})Q_{10} = 6.7\,\mathrm{nm}$ for $\Phi_{\mathrm{DOPE}} = 0.7$. In the middle curve both phases can be seen to coexist.

In conclusion, we have touched upon a fascinating field of soft (bio)matter that is readily accessible to x-ray investigations. In the meantime, theoretical work has pointed to a considerable variation of models for these types of phase.

[10] DOPE stands for di-oleoyl phosphatidylethanolamine.

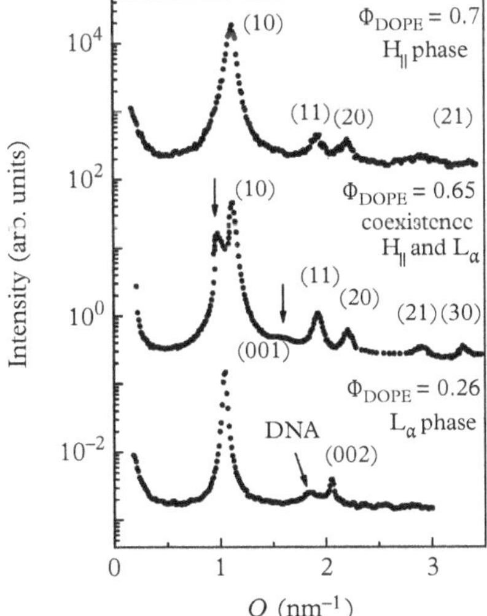

Figure 5.20 *Conversion of the lamellar L_α phase (bottom) to the inverted hexagonal phase H_{II} (top) as a function of the increasing weight fraction of DOPE, Φ_{DOPE}. In the middle graph both phases coexist: In addition to the hexagonal peaks, lamellar peaks can be seen as indicated by the arrows (after Safinya, 2001).*

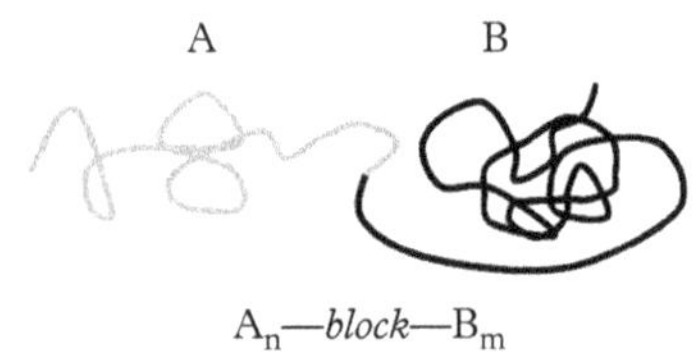

Figure 5.21 *Schematic representation of an AB diblock copolymer.*

5.2.5   Block copolymers

The simplest case of a block copolymer is a diblock consisting of two covalently bonded polymer blocks with chemically distinct repeat units A and B (Figure 5.21). As in principle polymers hardly mix,[11] this situation can be considered as the macromolecular equivalent of surfactants, discussed earlier. Added to a mixture of homopolymers A and B, AB blocks will go to the interfaces. Also, AB blocks can micro phase-separate into a whole series of structures like a spherical, cylindrical, or lamellar.[12] The phase behaviour depends on the interaction between the A and B units and their relative volume fraction. The interaction is described by χ_{AB}, the Flory-Huggins interaction parameter,[13] that is given by

$$\chi_{AB} = \frac{Z}{k_{\mathrm{B}} T} \left[\varepsilon_{\mathrm{AB}} - \frac{1}{2}(\varepsilon_{\mathrm{AA}} + \varepsilon_{\mathrm{BB}}) \right]. \qquad (5.1)$$

In this equation, ε is the interaction energy and Z the number of nearest-neighbour contacts. Note that $\chi \sim 1/T$. In the case of roughly equal block sizes, the block copolymer microphase separates into a lamellar phase. The lamellar period L scales as $L \sim N^{2/3}$, N being the number of monomers. This result stems from the balance between the enthalpy gain of demixing A and B (Equation 5.1) and the entropy cost of chain confinement within the layers. More generally the phase behaviour depends (apart from, of course, the molecular architecture) on the following parameters.

- The monomer interaction Equation (5.1) that is often empirically described by $\chi = \alpha T^{-1} + \beta$, in which α and β are experimentally determined enthalpy and excess entropy coefficients.

- N, the number of segments per polymer molecule.

- $f_{\mathrm{A}} \equiv \phi_{\mathrm{A}} = N_{\mathrm{A}}/(N_{\mathrm{A}} + N_{\mathrm{B}})$, the volume fraction of A-segments.

χN is the factor determining the phase behaviour, tailored via T and N. We can distinguish the following cases:

$\chi N \ll 10$ homogeneous state (entropy dominates).

$\chi N \approx 10$ order-disorder transition at T_{ODT}.

$\chi N \gg 10$ strong segregation.

For increasing deviations from about equal volume fractions ($f_{\mathrm{A}} > 0.5$ or $f_{\mathrm{A}} < 0.5$), the lamellar structure successively gives way to a hexagonal cylindrical and a spherical micellar structure, very similar as observed for low-mass surfactants (see Section 5.2.3).

A classical bulk phase diagram for the diblock copolymer system PI-*b*-PS[14] is shown in Figure 5.22. It was established in the 1990s by x-ray and neutron scattering. The types of phase encountered strongly resemble those discussed

[11] Unlike the mixing of simple fluids, for dissimilar polymers the entropy of mixing is small (varying inversely with molecular mass). Thus, even minor differences between A and B blocks tend to prevent mixing.

[12] It is assumed that the blocks are sufficiently uniform, that is, have a narrow molecular mass distribution.

[13] See, for example, Bates (1991).

[14] PI stands for polyisoprene

PS stands for polystyrene

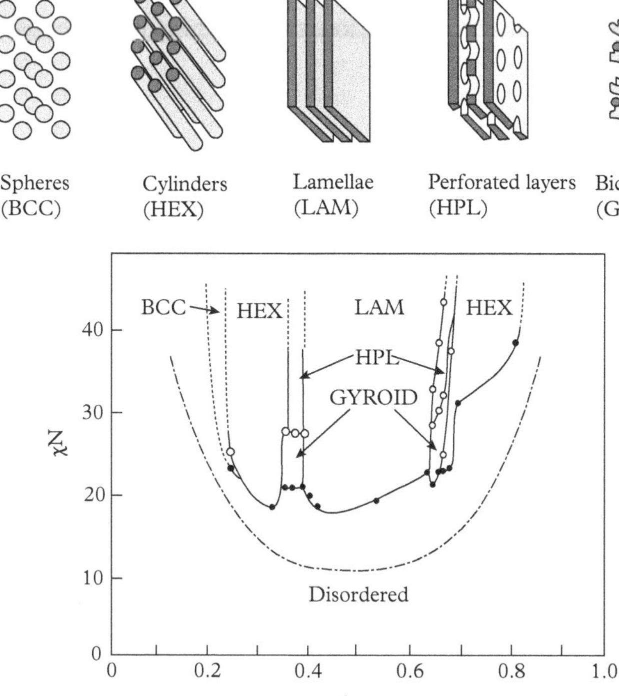

Figure 5.22 *Experimental phase diagram of the diblock copolymer PI-b-PS with on top schematic drawings of the various phases observed. (Adapted with permission from Khandpur et al., 1995. Copyright American Chemical Society.)*

for amphiphilic compounds. New is the hexagonally perforated lamellar phase (HPL) that is observed at the transition between the lamellar and bicontinuous phases.

For a further discussion of x-ray results we have chosen the system PBh-*b*-PEO[15] at different fractions and molecular mass (Lambreva, 2005). The x-ray pattern of PBh_{3700}-*b*-PEO_{4300}[16] is given in Figure 5.23; it indicates a lamellar phase, as expected for about equal volume fractions (in fact $f_{PBh} = 0.53$). We note

[15] PBh stands for hydrogenated polybutadiene, statistically distributed n : m = 1 : 1

PEO stands for polyethyleneoxide

[16] The subscript indicates the molar mass.

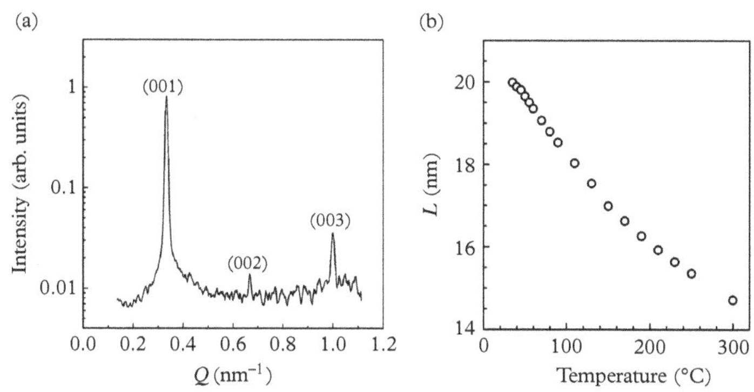

Figure 5.23 *SAXS results for PBh_{3700}-b-PEO_{4300} indicating a lamellar phase. No order–disorder transition (ODT) is observed.*

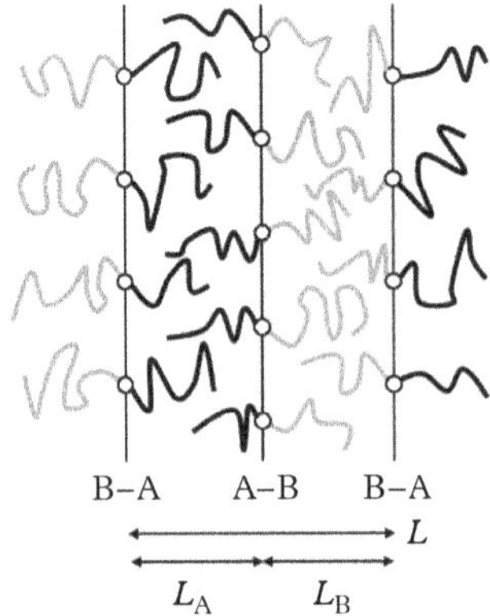

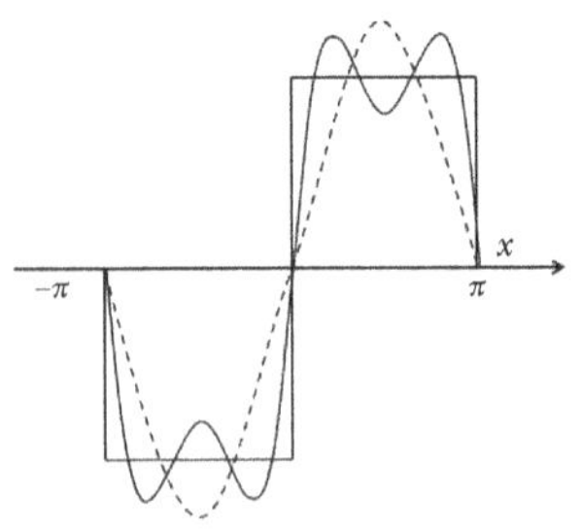

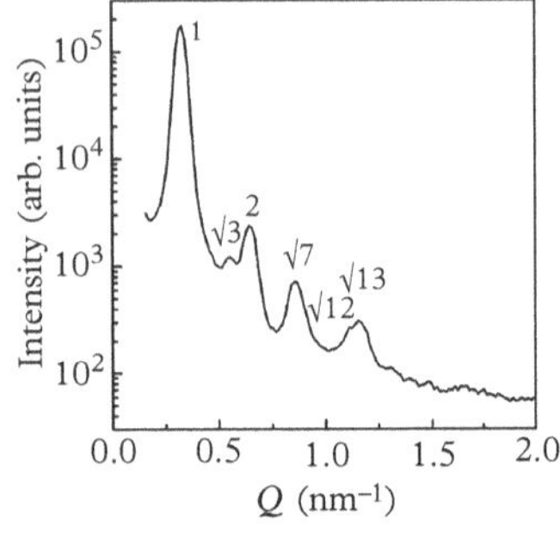

Figure 5.24 *Schematic picture of a symmetric lamellar diblock copolymer structure.*

Figure 5.25 *Approximating a square wave by* sin x *(broken line) and by the combination* sin x + $\frac{1}{3}$ sin $3x$ *(full line) indicates that (by symmetry) even terms like* sin $2x$ *do not contribute.*

Figure 5.26 *X-ray pattern from hexagonal cylinders for PBh$_{3700}$-b-PEO$_{2900}$ at 48°C (after Lambreva et al., 2005).*

a strong temperature dependence of the period L, but up to about 300°C there is no evidence for a transition to disorder. Remarkable is the rather weak second-order peak, which indicates sub-periods L_A and L_B that are about equal (compare with Figure 5.24). If the sub-periods L_A and L_B are exactly equal, the modulation along the layer normal can be described by a Fourier series of only odd terms (see Figure 5.25). In that case, all even harmonics in the x-ray pattern will be absent. The situation of Figure 5.23 with a small second-order peak indicates that the sub-periods L_A and L_B are close to each other. We conclude that the peak positions give information about the lamellar period while the intensities of the various orders allow us to determine the sub-periods of the individual blocks.

As a next step we reduce the molecular mass of the PEO block, arriving at PBh$_{3700}$-b-PEO$_{2900}$. This leads to f_{PBh} = 0.63 and brings the system to a hexagonal structure: see Figure 5.26. This can easily be concluded by comparing the ratios of the various peaks with Table 5.1. The basic hexagonal dimension is $2\pi/0.33$ = 19 nm.

To bring the order–disorder transition within reach, the molecular mass must be reduced, which has been done in the next example PB$_{1270}$-b-PEO$_{1450}$. In this case, f_{PBh} = 0.55 and we are again in the lamellar region. Figure 5.27a shows the first-order peak for increasing temperature. Between 110 and 112°C, the nature of the peak changes. This is hardly visible in the peak position, but according to Figure 5.27b and c is very clear in the intensity as well as in the width of the peak, leading to T_{ODT} = 110°C. A more detailed study of the lineshape (Figure 5.28) indicates a Gaussian below the ODT (long-range order, compare with Section 3.2) that changes into a Lorentzian shape above the ODT (short-range order).

Obviously, we can expand diblock systems to more blocks. The simplest situation is a triblock of the form A-B-A or B-A-B. Evidently this involves the same single parameter χ_{AB}, but the treatment of the volume fractions becomes more intricate. More interesting are A-B-C triblocks that can be linear or star shaped (see Figure 5.29). Now the combination of two composition variables (e.g., volume fractions f_A and f_B as f_C is not independent, $f_A + f_B + f_C$ = 1) and three χ parameters (χ_{AB}, χ_{AC}, and χ_{BC}) leads to a wide range of possible morphologies. In Figure 5.29 the simple case of a triblock with $f_A = f_B = f_C$ = 1/3 is illustrated, which allows for a linear and a three-arm structure. In practice, the wide parameter space leads to important variations in repulsions and attractions and gives rise to a zoo of different structures. Usually these are more complex than discussed so far for diblock systems. The various (in)compatibilities can lead to frustration and reaching equilibrium is more challenging than for diblock structures (Bates and Fredrickson, 1999). To get clear evidence for a particular morphology, SAXS data alone are often not conclusive and they have to be combined with TEM images.

Without going into great detail, we consider two examples of triblock systems. Bailey, Pham, and Bates (2001) extended the binary phase diagram of PI-b-PS of Figure 5.22 to a third dimension by attaching different percentages PEO as

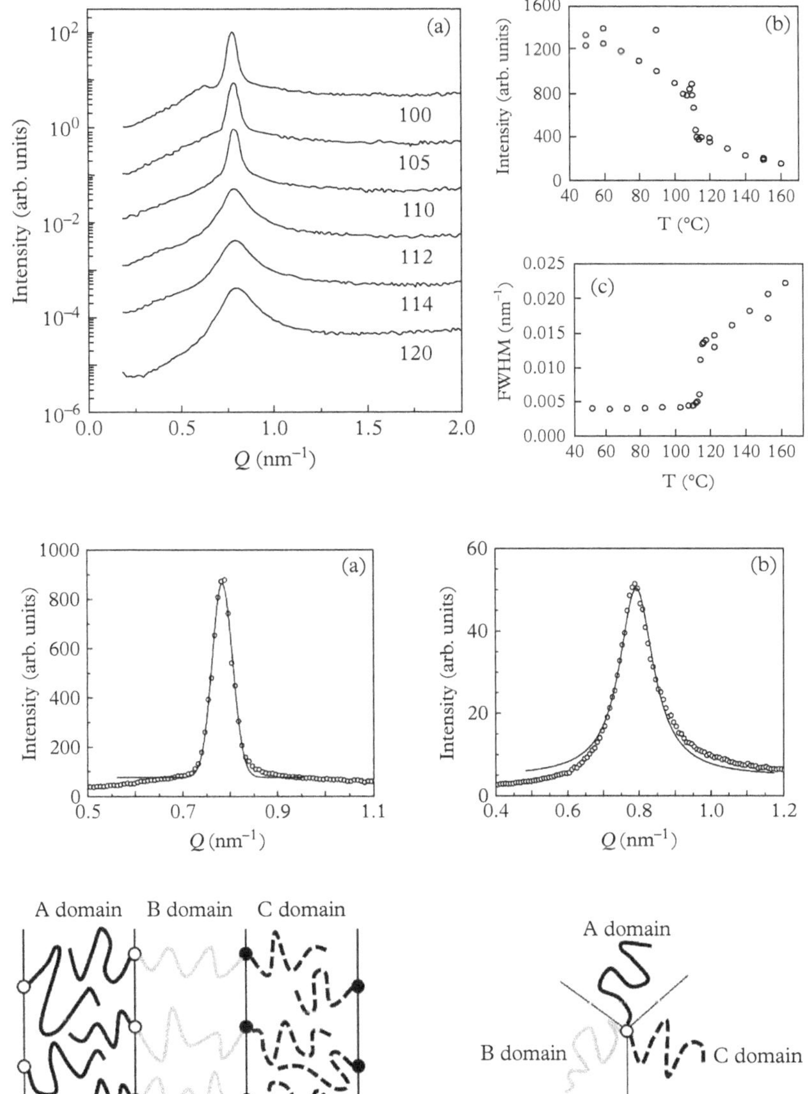

Figure 5.27 *X-ray results around the ODT of* PBh_{1270}-*b*-PEO_{1450}. *(a) X-ray peaks at the temperatures indicated. (b) Intensity, and (c) FWHM around the ODT (after Lambreva, 2005).*

Figure 5.28 *Variation of peak shapes in Figure 5.27a from (a) Gaussian below the ODT at 105°C to (b) Lorentzian above the ODT at 112°C.*[17]

Figure 5.29 *Topologies for a triblock copolymer. (a) Linear triblock. (b) Three-arm star triblock.*

a third component to always symmetric PI-*b*-PS compositions. In Figure 5.30 the simple case of a symmetric linear triblock with $f_A = f_B = f_C = 1/3$ is illustrated. The diffractogram of Figure 5.30a displays a fundamental peak and three higher orders at equal periods, evidently indicating a lamellar structure. It needs additional TEM photographs to identify a three domain structure of lamellae-*within*-lamellae. Similarly, the combination of SAXS and TEM of Figure 30b identifies convincingly core-shell cylinders for 16.6 vol % PEO.

[17] Another determination of the x-ray line shape around an ODT is given by Mai et al. (1996).

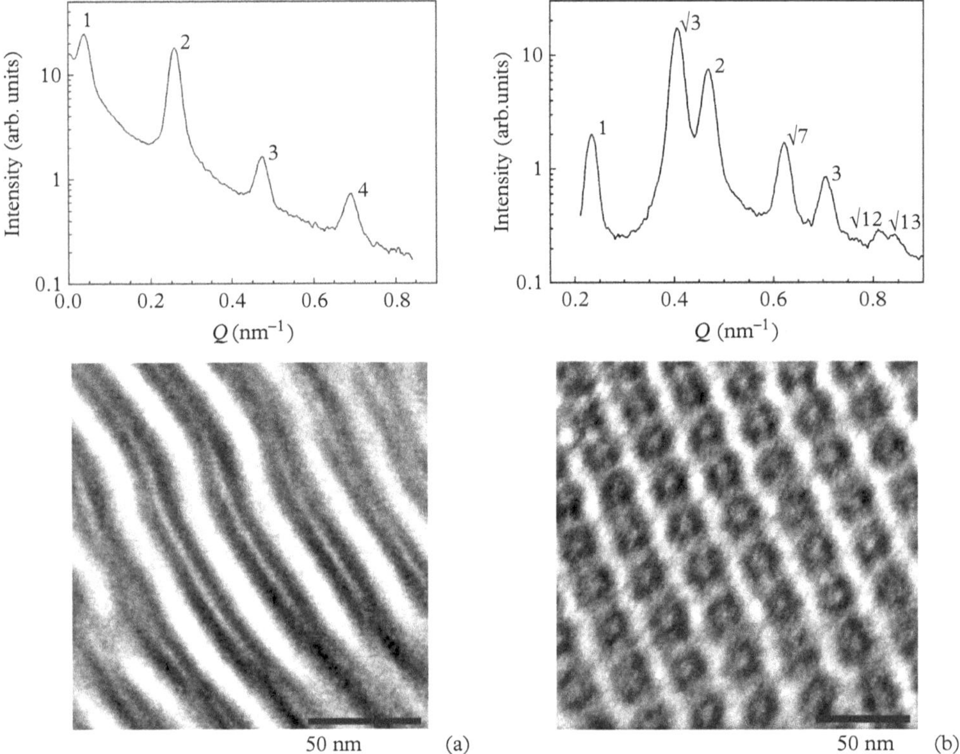

Figure 5.30 *(a) Three-domain lamellar structure at 145° C for symmetric PI-b-PS with 33 vol % PEO (compare with Figure 5.29a). In this situation, $f_{PI} = f_{PS} = f_{PEO} = 1/3$. (b) Core-shell cylinders for $f_{PEO} = 1/6$ (and thus $f_{PI} = f_{PS} = 5/12$) at 100° C. (Reprinted with permission from Bailey, Pham and Bates, 2001. Copyright American Chemical Society.)*

5.3 Case study: Bridging length scales—Order and frustration

In Sections 5.1 and 5.2, we considered various systems that give periodic structures (in one or more dimensions) that can be investigated with scattering methods. This case study regards a mixture of various aspects discussed so far (liquid crystallinity, amphiphilic behaviour, biomembranes, and blockcopolymers) and their associated length scales. The combination of different length scales is not always compatible with optimum spatial order, leading to frustration inside the system. This is a vast area of research we cannot do justice. We restrict ourselves to a couple of examples and refer the reader to the reviews by Förster and Konrad (2003) and by Ikkala, Houbenov, and Rannou (2014) for more.

5.3.1   Confined crystallization in block copolymers

As a starting point, we consider the simplest situation of order and frustration in a diblock copolymer: crystallization of one of the blocks that is confined by the interfaces of the other one. We return to the lamellar diblock system PBh_{3700}-b-PEO_{4300} of Figure 5.23. PEO is well-known to crystallize below the melting point of about 60°C into a monoclinic structure with $a = 0.81$ nm, $b = 1.31$ nm, $c = 1.96$ nm, and $\beta = 125°$. The different behaviour in a block-copolymer is not in the crystal structure of the PEO block (WAXS) but comes in with respect to SAXS of the long period (average stem length of the folded chain). In homopolymers the chain folds are metastable, induced by kinetic crystallization effects. In contrast, for block copolymers chain folding in the crystalline layer confined by the amorphous blocks can lead to an equilibrium degree of folds.

The lamellar diblock system PBh_{3700}-b-PEO_{4300} is strongly segregated with an ODT well above 300°C, and upon cooling from high temperatures we expect the PEO to crystallize below 60°C. The effect on the block morphology is shown in Figure 5.31 for a temperature jump from 100°C to 48°C.[18] Initially we observe the SAXS peak of the lamellar microphase separation at $Q \approx 0.32$ nm^{-1} ($d = 19.6$ nm). In addition to this fundamental peak we note a third-order one (compare with Figure 5.23). Evidently this defines a situation of alternating amorphous PEO and PBh lamellae in which the crystallization of PEO takes place. Only after about five minutes does a peak corresponding to crystallized PEO lamellae appears at smaller values ($Q \approx 0.25$ nm^{-1}). Hence, there is a clear thickening of the lamellae during the crystallization process. Moreover, in the crystallized state now, a weak second-order peak also becomes visible: the lamellar sublayers differ more in thickness than before crystallization. Evidently the original lamellar structure sets the length scale for the long spacing of the PEO crystals leading to a well-defined, relatively narrow peak.

Figure 5.32a shows the temperature dependence of long spacing associated with the lamellae before and after crystallization; each point corresponds to a specific crystallization temperature, T_{cr}. In the amorphous state the lamellar spacing

[18] In Figure 5.31 the SAXS curves are systematically shifted upwards. Hence the vertical axis also indicates time.

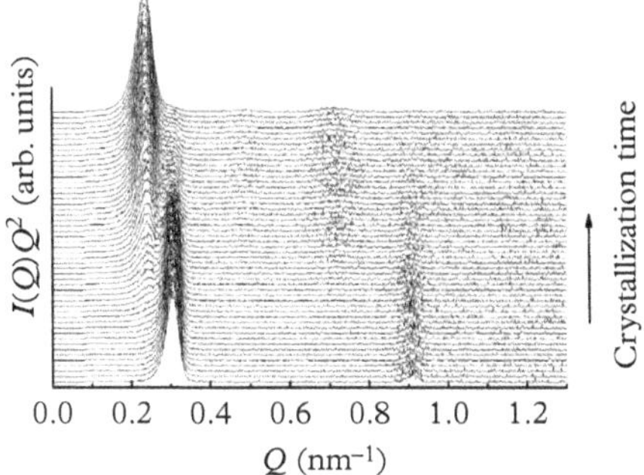

Figure 5.31 *SAXS measurements of isothermal crystallization of PBh_{3700}-b-PEO_{4300} after equilibrating at 100°C and cooling to 48°C at 30°C/min. Then measurements were started at 30 s/frame (after Li, Lambreva, and de Jeu, 2004).*

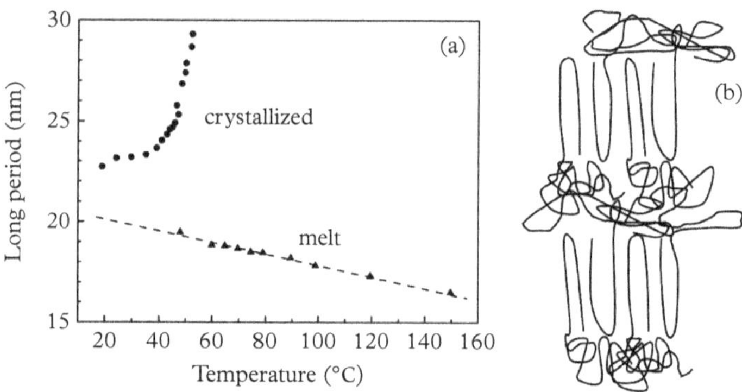

Figure 5.32 *(a) Long period of PBh$_{3700}$-b-PEO$_{4300}$, crystallized and in the melt (after Li, Lambreva, and de Jeu, 2004). (b) Cartoon model of once-folded crystalline PEO lamellae at 53°C.*

increases approximately linearly with decreasing temperature. This can be attributed to an increase of the Flory–Huggins parameter χ, which varies as $1/T$. The long period of the crystallized PEO lamellae increases continuously with T_{cr} (Figure 5.32a). Hence, integer-folded chain crystals cannot be assigned to the structures. The crystals with $L \approx 29.4$ nm obtained at 53°C are supposed to be equilibrium-folded chain crystals with approximately one fold, as in Figure 5.32b. According to the volume fraction of PEO, this would correspond to a crystal thickness of $0.445 \times 29.4 = 13.1$ nm.[19] Twice this value is in reasonable agreement with the extended chain length of 27.3 nm. Other crystals grown at lower temperatures must be still non-equilibrium crystals.

The situation of confined crystallization, as described, is relatively simple. However, if the values of T_{ODT} and T_{cr} become closer, interaction between the two processes (micro-phase separation and crystallization) occurs. For smaller values of T_{ODT} the phase segregation is weaker and the confinement of the crystallization process is not strong anymore. This can lead to a break-out of the crystalline structure, disrupting the previously existing melt microphase-separated morphology (Loo, Register, and Ryan, 2002). In intermediate cases the melt morphology is destroyed, but still influences the orientation of the crystalline lamellae (templated crystallization).

5.3.2 Liquid crystalline block copolymers

For more intricate combinations of length scales we take, as a starting point, comb-shaped liquid crystalline polymers and simple diblock copolymers. A cartoon model that combines the two structures is depicted in Figure 5.33. In this system the smectic order of the comb-shaped liquid-crystalline polymer can be considered as 'nanophase' separation. The polymeric backbone separates from the mesogenic units and can be qualitatively described by a (small) interaction parameter, χ_{BC}. In this model the system can be treated as a triblock copolymer. To illustrate the various possibilities for structure formation we consider two extreme situations.

[19] The volume fractions of the PEO block before and after crystallization are about 47% and 44.5%, respectively.

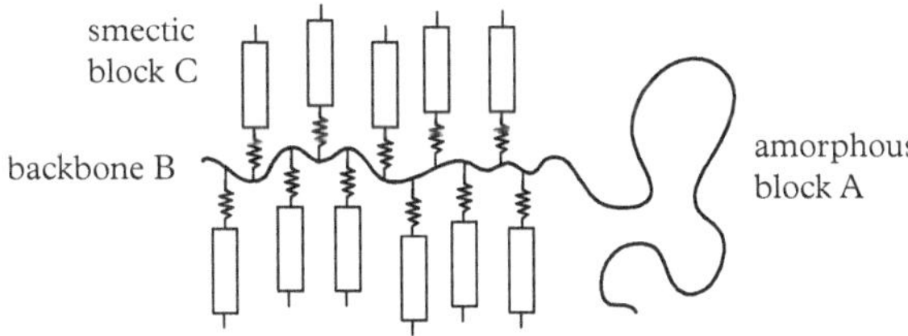

Figure 5.33 *Cartoon of a diblock copolymer with, on the left-hand side, a side-chain smectic polymer.*

Case 1. Let us assume that χ_{AB} is large. Then we have, effectively, an AB diblock system for which inside the B-block a smectic phase occurs. Upon increasing the temperature a smectic-isotropic phase transition will be encountered at T_{SI} (or alternatively a smectic-nematic one) that is hardly expected to influence the diblock system.

Case 2. Now we assume that the backbone of the liquid crystalline polymer is the same as the amorphous one: $\chi_{AB} = 0$. Then any block separation will be due to a finite χ_{AC}. In this situation the phase-separated block morphology can be expected to disappear when, upon heating, the smectic phase becomes unstable: T_{SI} equals T_{ODT}.

In hybrid amorphous/smectic block copolymers a hierarchical interplay arises between block copolymer ordering on the scale of tens of nm and smectic layering on the scale of nm. For strong block segregation the smectic phase is confined by the amorphous block. For weak segregation coupling between amorphous block and smectic structure becomes important.

As an example we consider the fluorinated smectic-amorphous diblock system of Figure 5.34a that consists of PMMA and a poly(acrylate) block bearing

[20] In fact, for compound F2 the smectic peak at room temperature is rather narrow, indicating in-plane order (CrB phase).

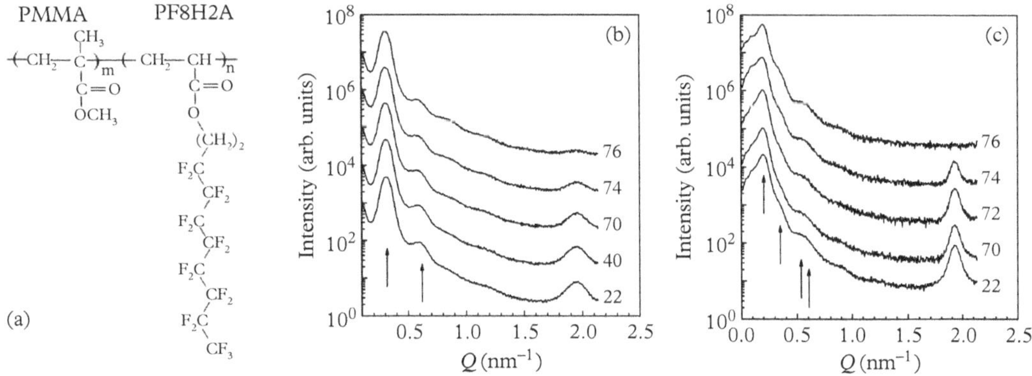

Figure 5.34 *(a) Smectic amorphous block copolymers F1 (m = 79, n = 394, and f_m = 0.354) and F2 (m = 80, n = 74, and f_m = 0.24). (b) Diffractogram of F1 with arrows at a ratio 1:2 indicating a lamellar phase. (c) Diffractogram of F2 with arrows at positions $1 : \sqrt{3} : \sqrt{7} : 3$ representative of a hexagonal phase (after Al-Hussein et al., 2005).*[20]

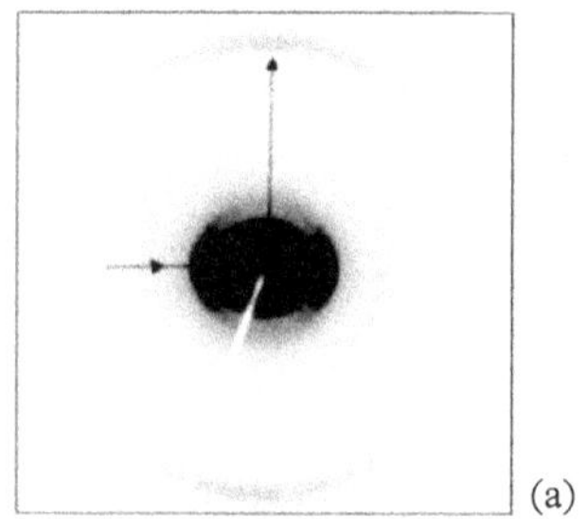 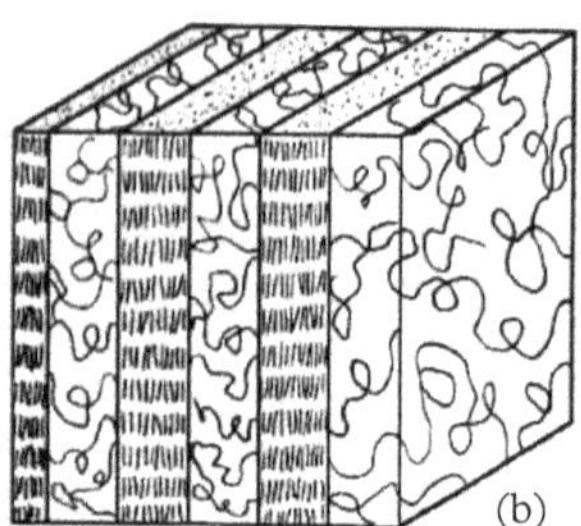

Figure 5.35 *(a) Two-dimensional x-ray picture of a uniformly ordered sample of compound F1. Inward and outward arrows indicate the reflections corresponding to microphase separation and smectic layering, respectively. (b) Model in which the smectic layers and block interfaces are orthogonal.*

semifluorinated alkyl side chains. The latter act as smectogenic liquid-crystal units while the length of the poly(MMA-acrylate) chain determines the morphology of the diblock separation. For compound F1 this is lamellar for the smaller PMMA block in F2 hexagonal.

In both cases the smectic phase is observed at around $Q \approx 1.9\,\mathrm{nm^{-1}}$ (3.3 nm). For F1, the lamellar block separation leads to x-ray peaks at around $Q \approx 0.21\,\mathrm{nm^{-1}}$, corresponding to 30 nm. Hence, we have a lamellar-*within*-lamellar structural hierarchy with two periods that differ by an order of magnitude. For both compounds the smectic layers are oriented perpendicular to the block interfaces, as illustrated in Figure 5.35 for F1.[21] For F2 we have a lamellar-*within*-hexagonal structure. The smectic phase gives way to an isotropic phase above 75°C which does not influence the x-ray signature of the block separation (see Case 1). The block segregation is strong with a high ODT that is not within reach.

5.3.3 Liquid crystalline dendrimers

The final example of frustration and order is at the intersection of dendrimers and liquid crystals. Dendrimers are uniform macromolecules with regular, highly branched three-dimensional structures. The number of successive branching points defines the generation number. Figure 5.36a shows a fifth generation carbosilane dendrimer. By their very nature dendrimers do not seem apt to further internal order. However, mesogenic groups can be covalently linked to the dendrimer's periphery by flexible aliphatic chains, effectively decoupling them from the relatively stiff dendritic core. The system to be discussed, as shown in Figure 5.36b, has mesogenic azobenzene groups at the periphery (Ostrovskii et al., 2013). We shall consider the fifth generation only, indicated as G-5-homo dendrimer. It displays a liquid crystalline morphology between its glass transition at −22°C and an isotropic phase above 62°C.

The small-angle x-ray peaks of the dendrimer are shown in Figure 5.37a, while in the wide-angle region a double-peaked diffuse profile is observed (Figure 5.37b). These results indicate a columnar phase, characterized by a two-dimensional in-plane lattice and liquid-like order in the perpendicular direction.

---

[21] Note the difference with the three-domain lamellar ABC block structure of Figure 5.30 in which the different lamellae are parallel.

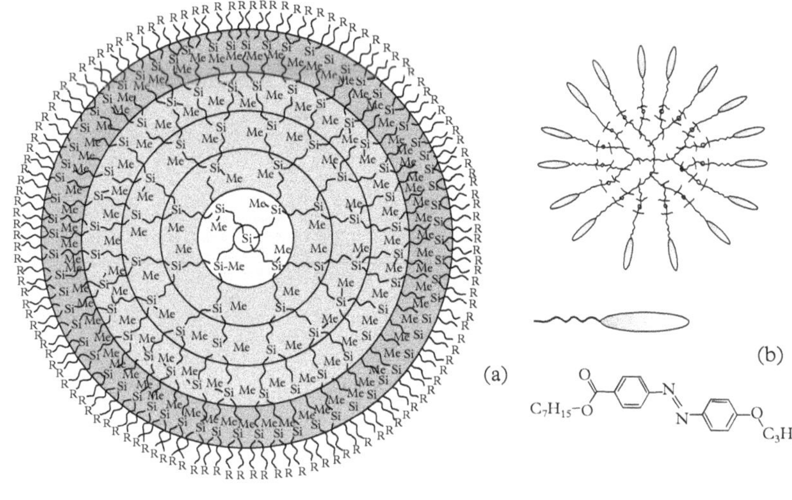

Figure 5.36 *(a) Example of a fifth generation dendrimer. (Reprinted with permission from Richardson et al., 2008. Copyright American Chemical Society.) (b) Scheme of the system under consideration (pictured for second generation only).*

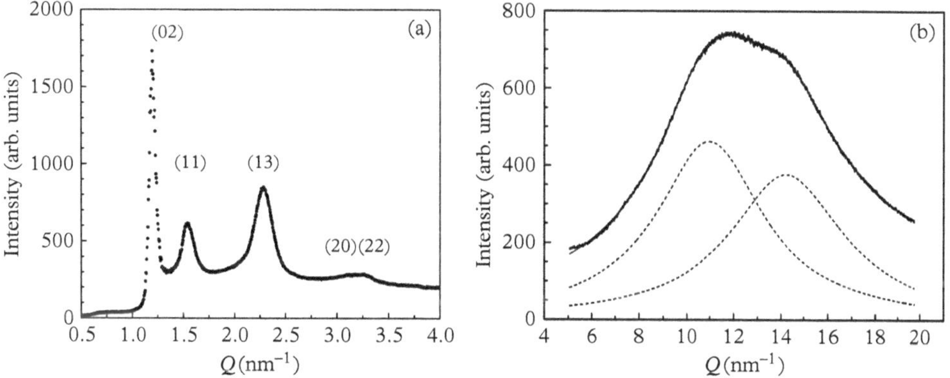

Figure 5.37 *X-ray results for the G-5-homo dendrimer at small angles (a) and wide angles (b), respectively. The latter peak can be decomposed into two Lorentzian ones.*

Though the highly symmetrical shape of LC dendritic molecules would suggest a hexagonal in-plane structure, this is not compatible with the observed sequence of peaks given in Table 5.2. Alternatively, a two-dimensional rectangular cell (simple rectangular or centred rectangular) must be assumed. For a centred rectangular cell (corresponding to the plane group *C2 mm*) extinction rules indicate that reflections satisfying $h + k$ = odd are forbidden (see Table 4.4 on page 67). From Table 5.2 we see that indeed only even combinations occur. Thus, most probably we are dealing with a centred two-dimensional lattice with unit-cell parameters $a = 4.39\,\mathrm{nm}$ and $b = 10.64\,\mathrm{nm}$. The lattice appears to be quite anisotropic with $b/a \approx 2.42$. This corresponds to a strong deviation from a standard hexagonal columnar phase for which $b/a = \sqrt{3}$.

Table 5.2 *Assignment of the x-ray peaks of Figure 5.37a.*

(hk)	Q_{exp} (nm^{-1})	d_{exp} (nm)	d_{calc} (nm)
02	1.18	5.322	5.32
11	1.55	4.058	4.06
13	2.28	2.752	2.75
20	2.83	2.224	2.22
22	3.11	2.019	2.02

Figure 5.38 *Schematic view of (a) the rectangular columnar phase of the G-5-homo dendrimer, with liquid-like order along the c-axis and ellipsoidal cross sections, (b) an individual molecule. (Reproduced from Ostrovskii et al., 2013, with kind permission from Springer Science and Business Media.)*

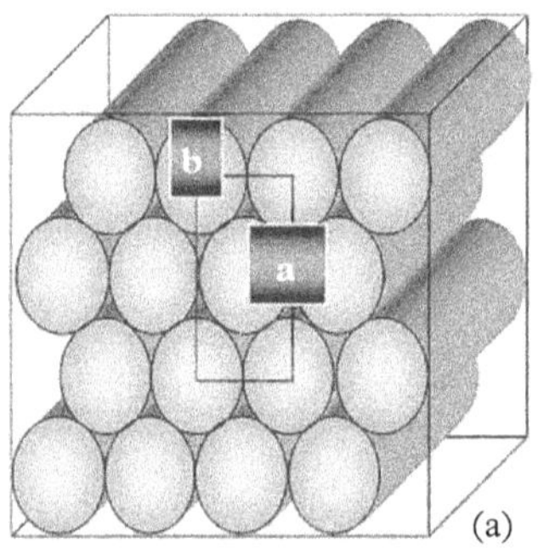
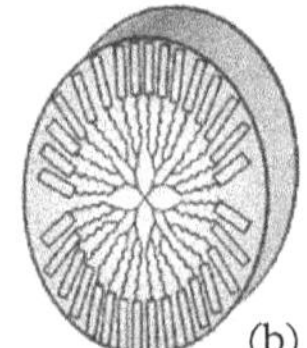

Perpendicular to the lattice plane the dendrimer molecules form columns that are internally disordered. The diffuse x-ray profile of Figure 5.37b can be well fitted by superposition of two Lorentzians centred at 11.02 nm^{-1} and 14.27 nm^{-1} (corresponding to 0.57 nm and 0.44 nm, respectively). The larger dimension relates to the liquid structure of the carbosilane fragments within the dendrimer matrix, while the second one corresponds to correlations of the flexible alkyl chains. The short-range in-plane order is characterized by correlation lengths of 0.7–0.9 nm, a few molecular diameters—quite normal for liquid order. Based on dense packing, elliptical molecular cross sections are assumed in the *ab*-plane and the overall picture is as given in Figure 5.38.

Soft Matter Films and Surfaces

6

In this chapter special attention is paid to the use of x-ray methods at interfaces and films of soft matter. Thin soft matter films are increasingly used in coatings and in many technological applications like liquid crystal displays and nanostructured block copolymer films. Furthermore, there is a fundamental interest in interfaces at which the structure and properties of a material interact with the outer world. First x-ray reflectivity (XRR), both at single and multiple interfaces, and its applications to several types of soft-matter films is discussed. Then methods of grazing incidence diffraction at wide angles (GID or GIWAXS) and small angles (GISAXS) are considered, including some examples. We finish with a case study in which a combination of the various techniques is used to elucidate the monolayer structure of some substituted thiophenes.

6.1 Reflectivity at a single interface

6.1.1 Fresnel theory

Consider reflection of x-rays against a single interface (Figure 6.1a).[1] In Section 1.6 we introduced the refractive index for x-rays as $n = 1 - \delta$, with δ small and positive, typically $\delta \approx 5 \times 10^{-6}$. Hence, the retractive index for x-rays is slightly smaller than one. Nevertheless, refraction and diffraction can in principle be described similarly. A more complete formula for n is given by

$$n = 1 - \delta + i\beta_{\mathrm{abs}} = 1 - \frac{\lambda^2}{2\pi}\mathrm{r_e}\rho + i\frac{\lambda}{4\pi}\mu. \tag{6.1}$$

This expression indicates that δ, and thus n, depends on the wavelength λ and the electron density ρ, while $\mathrm{r_e}$ is the Thompson scattering length of a single electron defined in Section 1.3. Furthermore, in Equation (6.1), absorption is included via the linear absorption coefficient μ (compare with Section 1.5).[2] Typical values for the coefficients are given in Table 6.1. For soft matter the value of n slightly smaller than one leads to a critical angle of total reflection, θ_{c},

[1] In the x-ray literature the angles are measured towards the surface and not, as usual in classical optics, with respect to the surface normal.

[2] For a more detailed discussion we refer the reader to Tolan (1999).

Table 6.1 *Scattering length densities* $r_e\rho$, *dispersions* δ, *linear absorption coefficients* μ, *and critical angles* θ_c *of some materials at* $\lambda = 0.1542$ *nm (CuK$_\alpha$ radiation).*

	$r_e\rho$ $(10^9\ \mathrm{mm^{-2}})$	$\delta(10^{-6})$	μ $(\mathrm{mm^{-1}})$	θ_c (deg)
Vacuum	0	0	0	0
PS $(C_8H_8)_n$	9.5	3.5	0.4	0.153
PMMA $(C_5H_8O_2)_n$	10.6	4.0	0.7	0.162
PVC $(C_2H_3Cl)_n$	12.1	4.6	8.6	0.174
PBrS $(C_8H_7Br)_n$	13.2	5.0	9.7	0.181
Quartz (SiO_2)	18.0–19.7	6.8–7.4	8.5	0.21–0.22
Silicon (Si)	20.0	7.6	14.1	0.223
Nickel (Ni)	72.6	27.4	40.7	0.424
Gold (Au)	131.5	49.6	417	0.570

The electron density has been defined in Box 2.1.

of typically $0.16°$.[3] Above this angle there is a transmitted beam at θ_t. (see Figure 6.1a). This diffraction at the interface can be described by Snell's law,[4] that can be written with $n_{air} = 1$ as

$$\cos\theta = n\cos\theta_t = (1-\delta)\cos\theta_t.$$

Total reflection occurs for $\theta_t = 0$, defining the critical (incoming) angle θ_c. Expanding the cosine we arrive at

$$\cos\theta_c \approx 1 - \frac{1}{2}\theta_c^2 + \ldots = n = 1 - \delta$$

[3] This is not different from the critical (Brewster) angle observed in classical optics at an interface glass/air, that is, however, much larger in magnitude.

[4] In France Snell's law is called Descartes law. However, the earliest accurate description of refraction is in the manuscript *On Burning Mirrors and Lenses* by Ibn Sahl (Baghdad, 984).

(a)

(b)

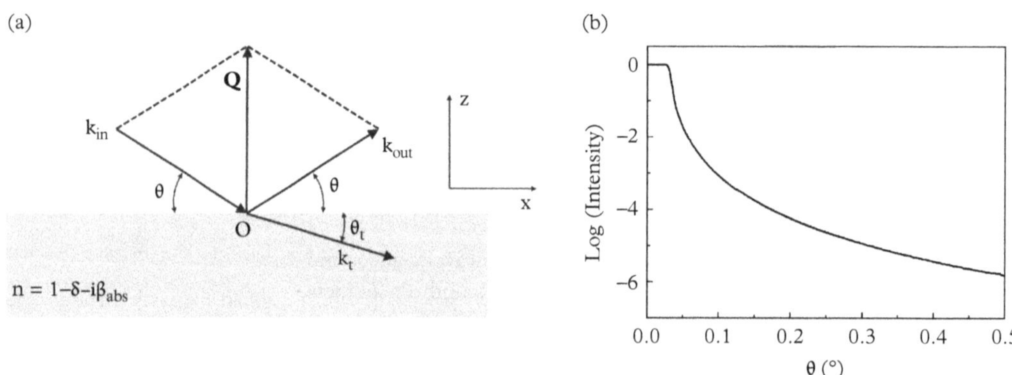

Figure 6.1 *(a) X-ray reflectivity against a single interface with (b) reflected intensity.*

or

$$\theta_c \approx \sqrt{2\delta} = \lambda\sqrt{r_e\rho/\pi}. \tag{6.2}$$

So far we have applied only secondary school optics to x-ray radiation. To describe x-ray reflectivity (XRR) we need a more complete theory.[5] Most importantly, this teaches us that above θ_c (the transmission regime) there is still some reflected intensity. However, its intensity decays steeply for $\theta > \theta_c$ (see Figure 6.1b). In fact the technique of XRR consists of measuring this reflected intensity over many decades, and is less concerned with the transmitted (refracted) beam. To become more quantitative, we need the basic result from electrodynamics that at the interface the incoming wave as well as its derivative must be continuous. This condition allows us to calculate the amplitude of the reflected as well as the transmitted wave. They are given by the so-called Fresnel coefficients expressed in terms of the scattered wave vectors $\mathbf{Q}$ and $\mathbf{Q}_t$ as

$$r(Q) = \frac{Q - Q_t}{Q + Q_t}, \quad t(Q) = \frac{2Q}{Q + Q_t}. \tag{6.3}$$

To describe XRR using these equations we need to eliminate the refracted wave vector Q_t.[6] From Figure 6.1 we note that for the specular situation $Q = 2k\sin\theta$ with $k = 2\pi/\lambda$. In the medium, $\lambda_t = \lambda/n$, and we find $Q_t = 2kn\sin\theta_t$. To obtain the relation between Q and Q_t we write Snell's law as

$$\cos^2\theta = n^2\cos^2\theta_t = n^2(1 - \sin^2\theta_t) \quad \text{or} \quad n^2\sin^2\theta_t = n^2 - \cos^2\theta.$$

Using $n^2 = \cos^2\theta_c$ (see Equation 6.2) this leads to

$$n^2\sin^2\theta_t = \cos^2\theta_c - \cos^2\theta = \sin^2\theta - \sin^2\theta_c,$$

or $Q_t^2 = Q^2 - Q_c^2$ giving $Q_t = \sqrt{Q^2 - Q_c^2}$, in which $Q_c = 2k\sin\theta_c$.

Now we can obtain the final results for the Fresnel coefficients (Equation 6.3) as

$$r(Q) = \frac{Q - Q_t}{Q + Q_t} \qquad\qquad t(Q) = \frac{2Q}{Q + Q_t}$$

$$= \frac{Q - \sqrt{Q^2 - Q_c^2}}{Q + \sqrt{Q^2 - Q_c^2}} \qquad\qquad = \frac{2Q}{Q + \sqrt{Q^2 - Q_c^2}}$$

$$= \frac{\sin\theta - \sqrt{\sin^2\theta - \sin^2\theta_c - 2i\beta_{\text{abs}}}}{\sin\theta + \sqrt{\sin^2\theta - \sin^2\theta_c - 2i\beta_{\text{abs}}}} \qquad\qquad = \frac{2\sin\theta}{\sin\theta + \sqrt{\sin^2\theta - \sin^2\theta_c - 2i\beta_{\text{abs}}}}$$

$$\tag{6.4}$$

The reflected intensity is calculated as $R_F = r\,r^*$, which reduces to $R_F = r^2$ when absorption is disregarded. Similarly, the transmitted one is written as $T_F = t\,t^*$.

[5] Such a complete theory of optics was given by Fresnel in the nineteenth century.

[6] Note that, although we will use quite some formulae in the following lines, in order to understand things, you do not need more than Equation (6.3) from Fresnel and simple optics as described by Snell's law.

Figure 6.2a shows the result for R_F around the critical angle for various values of β_{abs}; evidently absorption is only important in this region. Keeping this in mind we shall neglect absorption from now on unless explicitly stated otherwise.[7]

A useful approximation can be obtained by expanding Equation (6.4) for $Q > Q_c$ in terms of the small parameter Q_c/Q. For the Fresnel reflectivity this leads to

$$R_F = \left(\frac{Q_c}{2Q} \right)^4. \tag{6.5}$$

Figure 6.2b shows R_F over a wide Q-range with, in addition, the result of the approximate Equation (6.5). This expansion is perfectly valid for about $Q > 3Q_c$ for which $R_F \sim Q^{-4}$.

As we have seen, R_F can be calculated exactly. Information on the system (surface) investigated shows up as deviations from R_F of the measured reflected intensity $R(Q)$. Therefore it is best to present experimental reflectivity data as plots of $R(Q)/R_F$ vs. Q.

6.1.2 Practical aspects: Scanning, footprint, and roughness

Figure 6.3a illustrates the practice of performing a reflectivity scan. Two angles have to be changed: The incoming angle between the x-ray beam and the sample are moved at half the value of the scattering angle 2θ, leading to a variation of $Q = Q_z$ along the normal to the interface. In addition, at each point along the resulting reflectivity curve (fixed 2θ), the sample can be 'rocked' by varying $\omega = \alpha - \beta$. Though the **Q**-vector describes a curve in reciprocal space (see Figure 6.3b), for small angles the variations in Q_z are minor and effectively Q_x is probed.

The next point to be considered is the relation between the incident x-ray beam and the size of the sample. As indicated in Figure 6.4, an x-ray beam of width b projects on the sample surface with a length $L = b/\sin\theta$, the so-called footprint. Let θ_F be the angle for which the footprint is equal to the sample dimension along

[7] Formally these expressions are only valid for an incident wave polarized parallel to the surface. However, in practice, polarization effects are minor at the small angles involved for x-ray reflectivity.

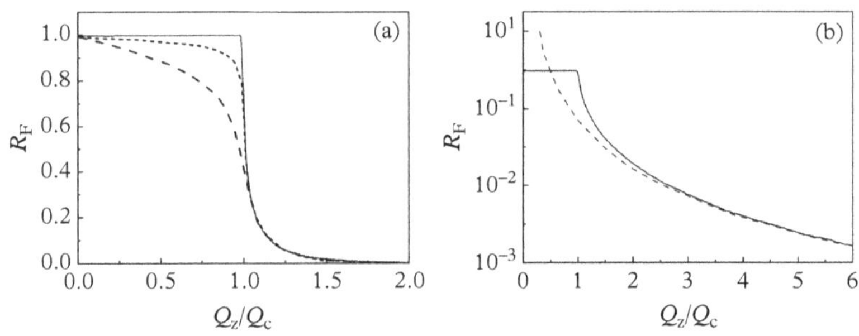

Figure 6.2 *Full lines: Fresnel reflectivity* R_F *without incorporation of absorption. (a) Broken lines indicate the effect of including absorption for* $\beta/\delta = 0.02$ *and* 0.1. *(b) Broken line shows the approximate formula, Equation (6.5).*

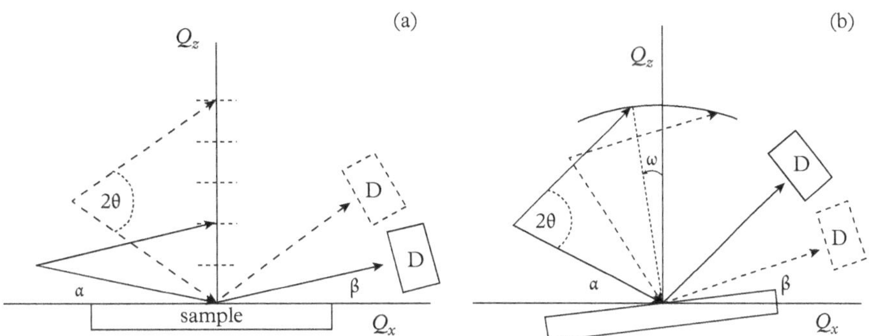

Figure 6.3 *(a) Reflectivity scan in which Q_z is varied by moving $\alpha = \beta$ at half 2θ. (b) Rocking curve in which $\omega = \alpha - \beta$ is moved at constant 2θ. At small angles, for this scan, mainly Q_x is varied. D represents the detector.*

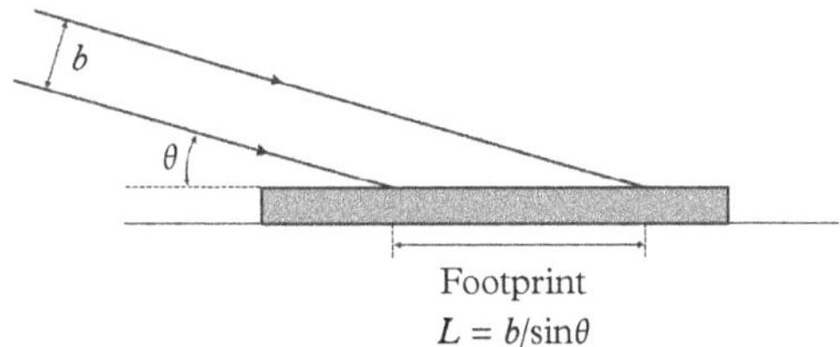

Footprint
$L = b/\sin\theta$

Figure 6.4 *Definition of the footprint of the beam along the sample.*

the beam. Then for $\theta < \theta_F$ the beam overflows the sample and the experimentally measured $R(Q)$ is smaller than the ideal one. It is customary to incorporate this effect by correcting the measured data for $\theta < \theta_F$,

$$R^{\mathrm{corr}} = \frac{\sin\theta_F}{\sin\theta}R^{\mathrm{exp}} \approx \frac{\theta_F}{\theta}R^{\mathrm{exp}}.$$

So far we assumed all interfaces to be nicely flat, which is not the case in reality. How do we incorporate any roughness of the interface? This is a delicate problem that depends on the scale investigated, as illustrated in Figure 6.5. For small roughness scales the average picture (a) applies. We can use an average interface but part of the specular reflected intensity is lost as diffuse scattering. On the larger length scale of (b) (large compared to the width b of the incident beam) we cannot use an average interface anymore. This situation is better described as a variable incident angle. A description of the transition region is, of course, rather intricate. Restricting ourselves to the first situation, the average density profile (as projected along the surface normal) will still be smooth. It can be approximated by an error function (see Figure 6.5c). In that case the derivative of the profile (a Gaussian function of width σ) is a measure of the roughness. The expression

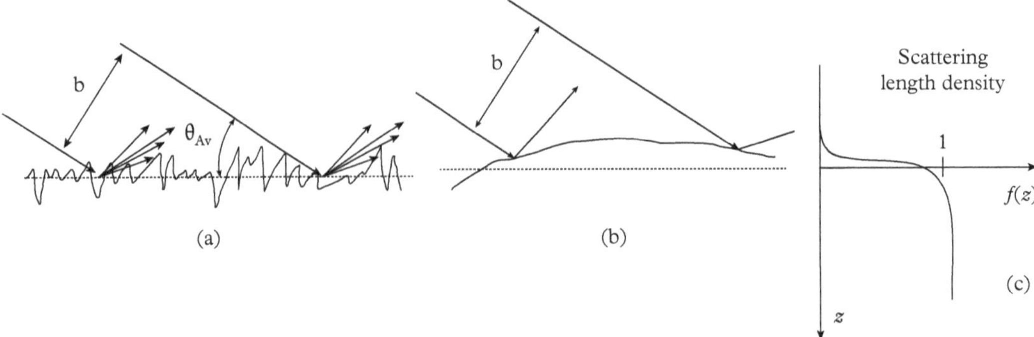

Figure 6.5 *Roughness on different length scales (a, b) can give the same average profile along the surface normal (c).*

for the reflectivity coefficient at the interface (Equation 6.3) can be extended to incorporate this effect,

$$r(Q) = \frac{Q - Q_t}{Q + Q_t} \exp\left(-\frac{1}{2} Q Q_t \sigma^2\right). \tag{6.6}$$

A more complete description of roughness requires knowledge of the height–height correlation function along the surface that involves the lateral distances, which is beyond our present scope. We still present Figure 6.6 to illustrate that very different height–height correlations can give the same Gaussian root-mean-square (rms) roughness. This short discussion should make clear that roughness values derived from XRR using Equation (6.6) and from, say, atomic force microscopy, are not necessarily similar because a different scale is probed.

6.1.3 Examples of soft interfaces

As a first example we consider a perfectly smooth liquid/air interface. Figure 6.7 gives XRR results for water over a large dynamic range. At large Q-values the

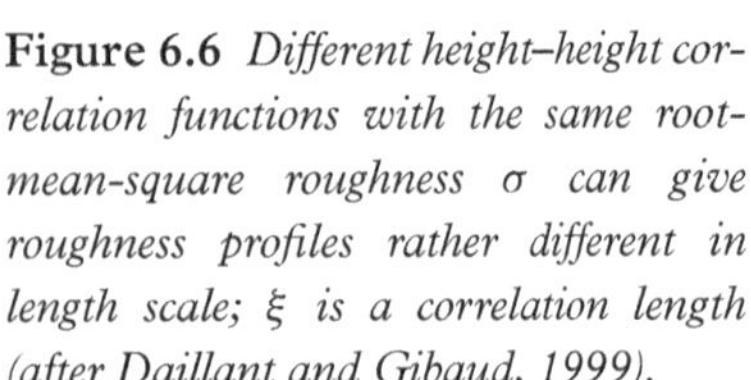

Figure 6.6 *Different height–height correlation functions with the same root-mean-square roughness σ can give roughness profiles rather different in length scale; ξ is a correlation length (after Daillant and Gibaud, 1999).*

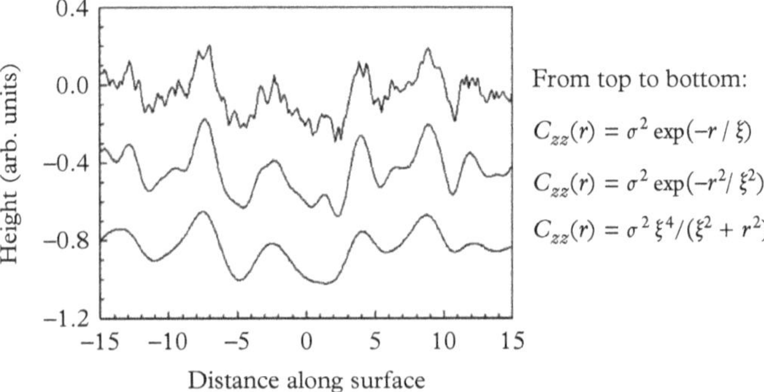

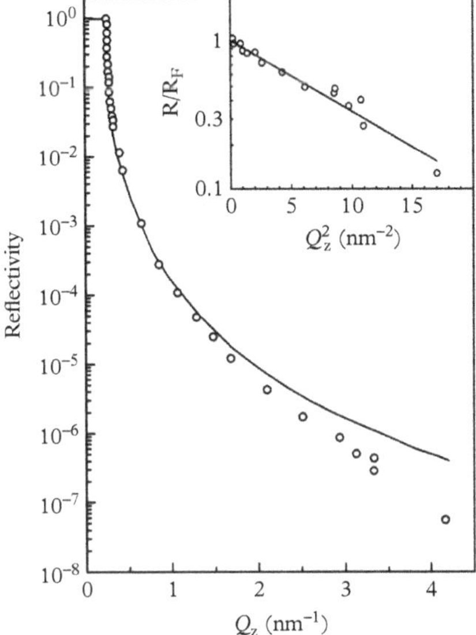

Figure 6.7 *Reflectivity R of a water/air interface (circles). The solid line represents R_F of a perfectly sharp interface. The insert indicates a linear variation of R/R_F with Q^2. (Reprinted with permission from Braslau et al., 1985. Copyright American Physical Society.)*

measured reflectivity falls well below R_F. This has been attributed to (dynamic) roughness due to thermally-induced capillary waves. As a result some intensity is scattered away from the specular direction (compare with Figure 6.5a) resulting in a lower specular intensity. The inset shows that a plot of $\log(R/R_F)$ varies linearly with Q^2, indicating that the roughness can be modelled by a factor $\exp(-\sigma^2 Q^2)$, in agreement with Equation (6.6) and the theory of capillary waves (Penfold, 2001).

The second example is a strongly polar liquid-crystalline compound at the nematic/air interface. XRR results are shown in Figure 6.8 around a nematic–smectic-A phase transition at T_{NA}. As discussed in Section 5.2.1 (Figure 5.8), upon approaching a smectic-A phase by lowering the temperature, in the bulk nematic phase pre-transitional smectic 'droplets' can be formed. In such a situation, a surface serves as a nucleation centre of smectic order with the layers parallel to the interface. The correlation length $\xi_{//}$ is the same as the longitudinal correlation length observed in the bulk (Figure 6.8b). As a result, in XRR a smectic layer peak is found at $Q = Q_0$ (Figure 6.8a) that increases in intensity upon approaching T_{NA} and subsequently saturates below the transition temperature. This is very similar to the situation observed for $\xi_{//}$ in the bulk but due to the interface $\xi_\perp$ is immediately very large.

Interestingly in Figure 6.8 another feature is observed at $Q = Q_0/2$ that—because of its Q-value—must be associated with double layers. This has been attributed to the strong end dipoles that are preferably oriented inwards at the interface (the direction of the polar liquid). This triggers off an anti-parallel dipole

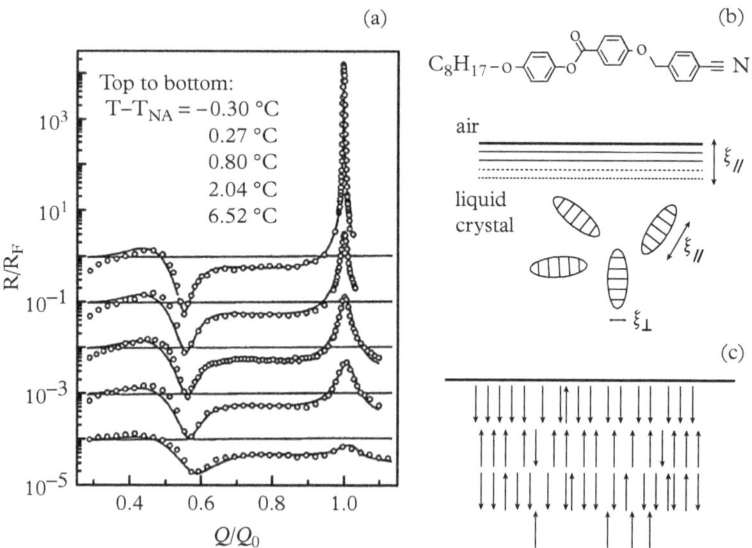

Figure 6.8 *XRR of the liquid crystal surface of a strongly polar compound close to the transition nematic to smectic-A at T_{NA} = 120°C (after Gramsbergen, de Jeu, and Als-Nielsen, 1986). (a) XRR indicating a smectic peak at Q_0 = $2\pi/L$ and an additional feature at $Q_0/2$. (b, c) Schematic representation of pre-transitional smectic order near the interface.*

orientation in the next layer, etc. (Figure 6.8c). However, there is a fundamental difference between the two types of effect. The smectic layer order at Q_0 develops in the nematic phase gradually, with decreasing temperature, and upon approaching the transition to the smectic phase $\xi_{//} \to \infty$. The anti-ferroelectric surface ordering is short-range with a finite size and occurs in both the nematic and smectic phase with a limited dependence on temperature.

6.2 Multiple interfaces: Soft matter films

6.2.1 Uniform films

Figure 6.9 extends the reasoning used so far to a single layer on top of a substrate in air ($n_0 = 1$). Using Equation (6.3) for the two relevant interfaces (writing r_{01}, t_{01} and r_{12}, t_{12}), we see easily that $r = r_{01} + t_{01}r_{12}t_{10}\exp(iq_1L)$ + multiple bounces. Making the summation over all multiple reflections leads to

$$r = \frac{r_{01} + r_{12}\exp(iQ_1L)}{1 + r_{01}r_{12}\exp(iQ_1L)}. \tag{6.7}$$

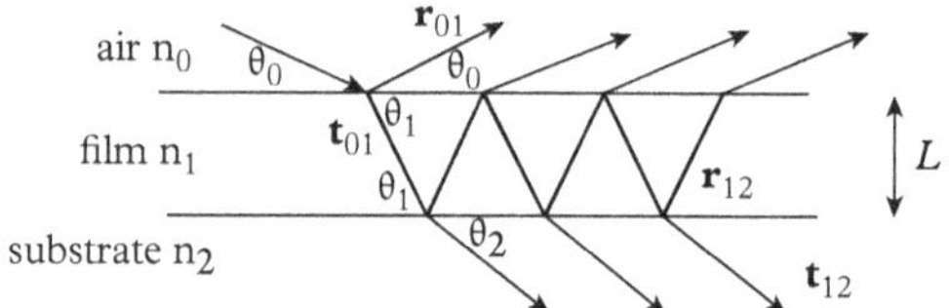

Figure 6.9 *Schematic representation of multiple reflections from a smooth thin film of thickness L.*

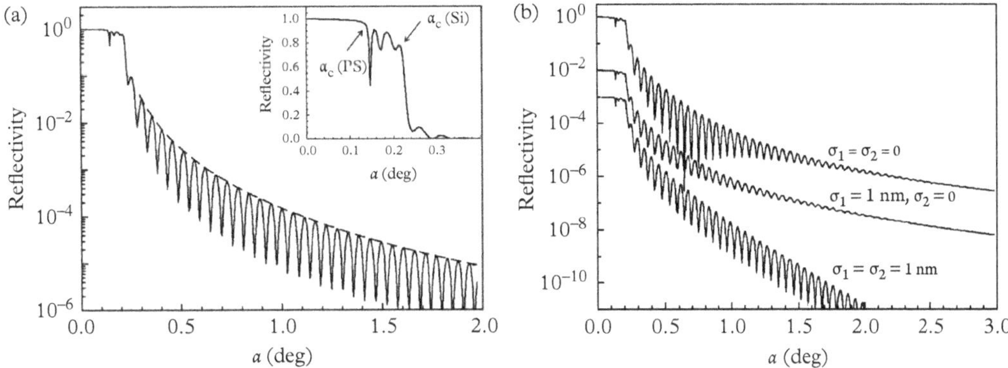

Figure 6.10 *Calculated reflectivity for a 80 nm polystyrene film on a Si substrate ($\delta_{PS} = 3.5 \times 10^{-6}$ and $\delta_{Si} = 7.56 \times 10^{-6}$). (a) Without roughness, with the Fresnel reflectivity of the bare substrate as the dashed line. (b) The effect of including roughness of each of the interfaces (σ_1 for PS-air, σ_2 for Si-PS).*

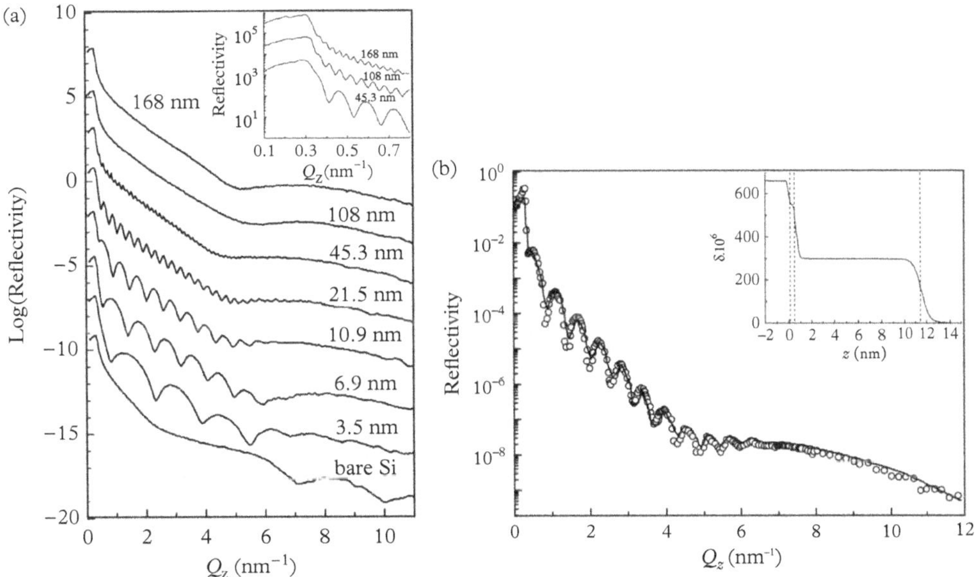

Figure 6.11 *(a) XRR results for polystyrene films of different thickness on a Si substrate. (b) Details of the 108 nm film with a fit to the density profile in the inset (after Tolan, 1999).*

Figure 6.10a shows the reflectivity calculated for an 80 nm uniform PS film in vacuum showing sharp oscillations—the so-called Kiessig fringes—which can be understood from Equation (6.7). The phase difference between the reflections from the two interfaces is $\exp(-iQ_1 L)$: constructive interference of x-rays reflected from the top and bottom of the film occurs with a period $2\pi/L$. The amplitude of these oscillations depends on the electron density contrast between film and surroundings and is at maximum when no substrate is present. The film thickness follows from the spacing between two maxima,[8] $\Delta\theta$ or ΔQ, using

$$L = \frac{2\pi}{\Delta Q} = \frac{\lambda}{2\Delta\theta}.$$

(6.8)

Experimentally the period should be preferably averaged over 6–10 maxima starting at not too small angles, typically $> 4\theta_c$. According to Equation 6.2) the critical angle θ_c provides a direct measure of the average density in the film.

Figure 6.10b shows calculations for a 80 nm PS film with and without roughness. As discussed, roughness causes some x-rays to be scattered away from the specular direction. Note that we can distinguish the roughness from each of the interfaces. Figure 6.11 displays a series of actual measurements of a PS film on a Si substrate. Note the effect of the SiO_x top layer on the substrate (bottom line of Figure 6.11a). Figure 6.11b also indicates the density profile used for a full fit to the data of one of the films.

6.2.2 Multilayers: Recursive formalism

The formulation used so far can easily be extended to an arbitrary number of slabs. The principle is shown in Figure 6.12a for a film with a periodic internal structure. As a result, on top of and interfering with the Kiessig fringes, Bragg-like peaks are observed. Remember that the width of a Bragg peak increases if the number of periods is small (Figure 3.4). Hence for films with just a couple of layers the interference with the Kiessig fringes blurs the Bragg peak as such. Figure 6.12b,c display a calculation and an experiment for a smectic film in vacuum. In both cases the periodicity is very similar ($Q_0 \approx 2.1-2.2$ nm^{-1}). However, the number of layers is larger for the experiment, leading to closer-spaced fringes.

[8] Hence XRR serves as a ruler with sub-nm precision (the wavelength of the x-rays used).

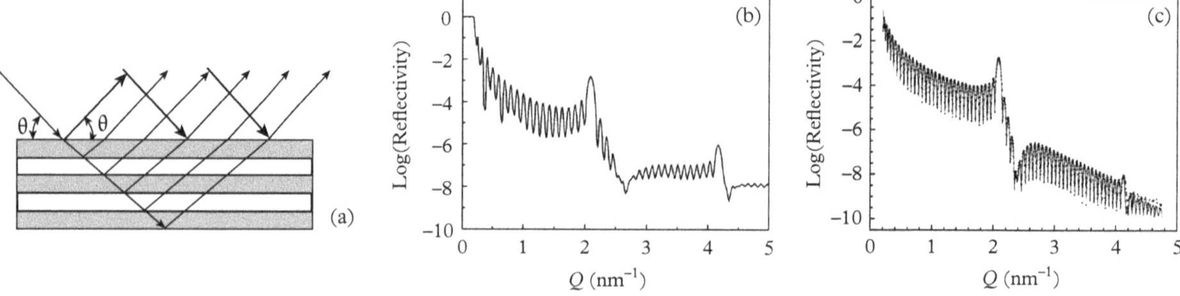

Figure 6.12 *(a) Model for a film of alternating layers. (b) Calculated reflectivity for a smooth 90 nm thick film in vacuum with an internal structure of 20 alternating layers of 3 nm with $\delta_1 = 3 \times 10^{-6}$, $\delta_2 = 2 \times 10^{-6}$, and $\beta = 10^{-8}$. (c) Experimental reflectivity from a smectic film of the compound 7AB (see Figure 3.5) with 33 layers of 2.87 nm.*

More formally, Figure 6.9 can be extended to the general multilayer system of Figure 6.13. To obtain the resulting reflectivity a recursive model is used (Parratt formalism). In practice various levels of approximation can be found in the literature and it is relevant to realize the approximations involved. These are illustrated in Figure 6.14. Case (a) is the simplest approach indicated as kinematical theory.[9] Obviously, the full dynamical theory of (c) is the most appropriate choice. The results of a full XRR measurement and its fitting require a model that contains (1) a number of (sub)layers, with (2) their electron density, and (3) the roughness of the interfaces. Computer programs are available to apply this formalism to a large number of interfaces.[10] An example is the program StochFit (Danauskas et al., 2008). Note that the resulting model is not necessarily unique. However, often circumstantial information is available for the system under investigation that helps to discriminate between different possibilities. Furthermore, it is important to realize that only information along the film normal (Q_z) is obtained. Hence a

[9] Also called the first Born approximation.

[10] Various methods and approximations to calculate XRR were compared by Van der Lee, Salah, and Harzallah, 2007.

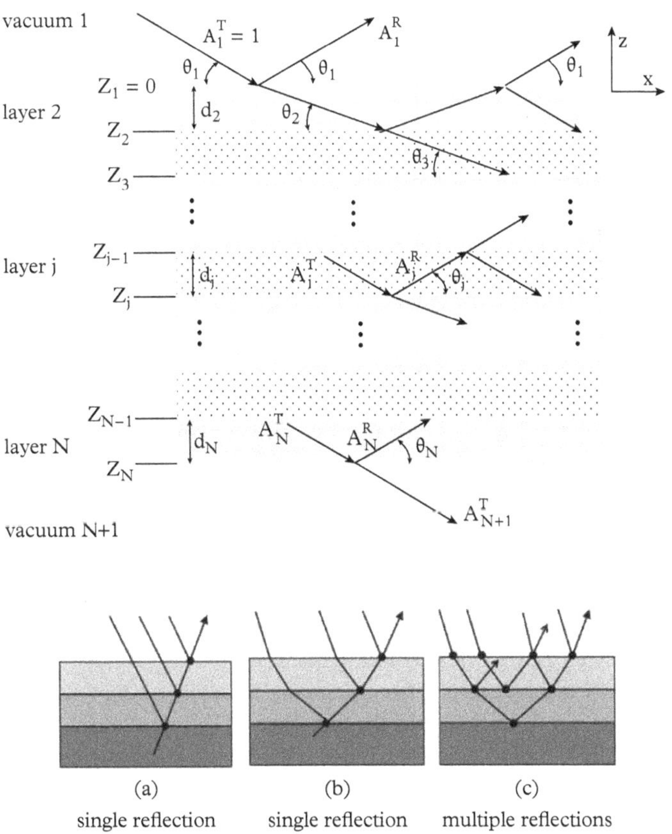

Figure 6.13 *Scheme to calculate the total reflectivity of a stack of layers.*

Figure 6.14 *Overview of the various theoretical approaches for a reflectivity calculation and the approximations involved.*

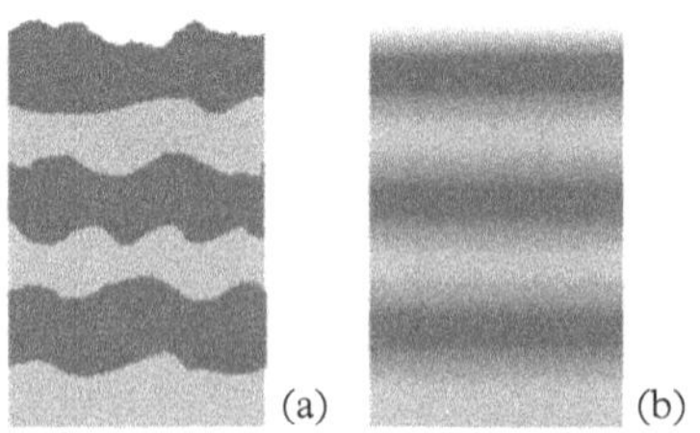

Figure 6.15 *A situation of rough layers as pictured in (a) is modelled in XRR with the density projected along the layer normal as in (b).*

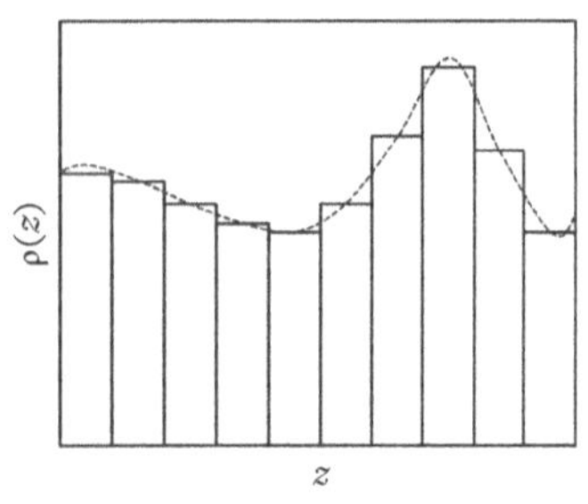

Figure 6.16 *A continuous density profile (dashed line) can be approximated by a number of slabs with constant density.*

layered system, as pictured in Figure 6.15a, will be 'seen' in XRR, as displayed in Figure 6.15b. To obtain lateral information grazing incidence diffraction can be used, see Section 6.3 below.

Finally we note that an arbitrary density profile can be approximated by a large number of slabs (see Figure 6.16). Neglecting refraction and multiple scattering (kinematical theory of Figure 6.14a) and taking the limit of an infinite number of infinitesimally thin slabs, one can derive the so-called master formula for specular reflectivity,

$$I(Q_z)/R_F = \frac{1}{\bar\rho}\left| \int_{-\infty}^{\infty} \frac{d\rho(z)}{dz} \exp(iQ_z z)\,dz \right|^2, \tag{6.9}$$

where $\bar\rho$ is the average density. The kinematical approximation is valid in the limit of weak interactions and diverges in the regime of total reflection. This can be corrected by replacing Q in the exponent of Equation (6.9) by its average value $\sqrt{Q^2 - Q_c^2}$ in the film under consideration. In this way, refraction corrections are still neglected for individual slabs, but the major refraction at the air/film interface is taken into account.

6.2.3 Example 1: Monolayers of fluorinated alkanes

Fluorinated alkanes like $CF_3(CF_2)_n(CH_2)_mCH_3$ are called primitive surfactants because of the difference between the fluorinated chain and the aliphatic tail. Like more conventional surfactants they easily form a monolayer at an interface. The difference in properties of the two types of chain can be summarized as follows:

$(CF_2)_n$: rigid 15/7 helical conformation; surface active (low surface tension).

$(CH_2)_m$: flexible zig-zag conformation; less surface active.

The Van der Waals diameters of a CF_2 and a CH_2 group are 0.60 and 0.49 nm, respectively, with corresponding cross-sections of 0.28 and 0.19 nm^2, which evidently must create interesting packing problems. As an example, we shall consider monolayers of $CF_3(CF_2)_{13}(CH_2)_{19}CH_3$ (in short notation $F_{14}H_{20}$) on a Si-substrate (Mourran et al., 2005). A comprehensive review of fluoro-alkane systems has been given by Krafft and Riess (2009).

Figure 6.17 displays results from scanning force microscopy that indicate—depending on the solvent used—the presence of either elongated ribbons or strongly curved ribbons leading to nano-spirals. XRR measurements give remarkably clear results (Figure 6.18). The pronounced minima indicate a well-defined thickness of 3.61 nm. The length of the extended molecule is 4.65 nm, which exceeds the layer thickness considerably. It is, however, consistent with an

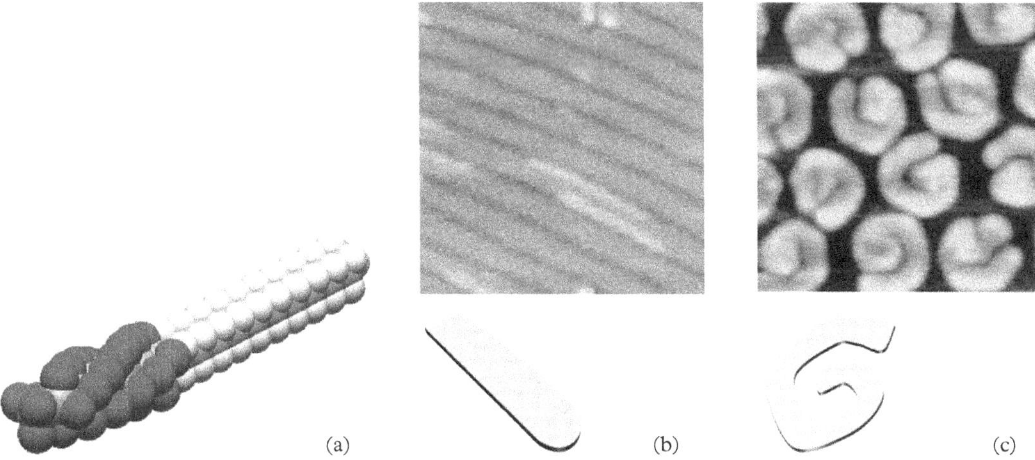

Figure 6.17 *(a) Model of the structure of the fluorinated alkane $F_{14}H_{20}$. (b, c) Scanning force microscopy (0.25 × 0.25 μm²) of monolayers of $F_{14}H_{20}$, deposited from a specific solvent and a neutral solvent, respectively. Below the picture cartoons are given of the resulting ribbon and spiral structure.*

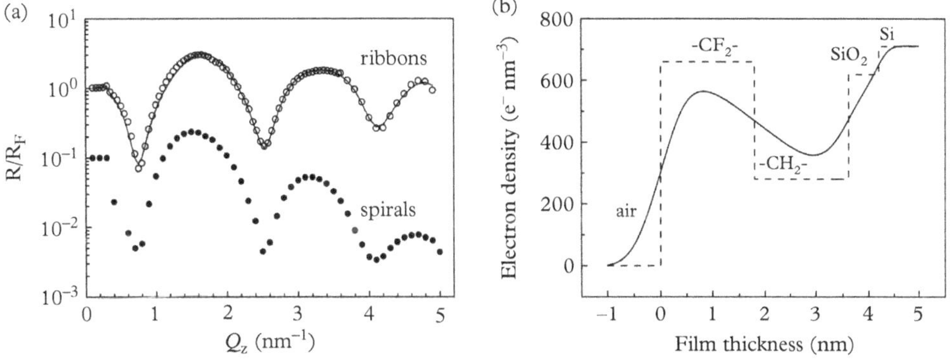

Figure 6.18 *XRR data of monolayer films of $F_{14}H_{20}$. (a) Normalized reflectivity for ribbons and spirals, respectively, shifted by one decade for clarity. (b) Electron density profiles from a fit to the ribbon data with and without roughness included (full and broken line, respectively).*

arrangement where the fluorinated chains are oriented normal to the surface layer and the alkyl segments tilted with—on average—a 122° angle between the two segments (see Figure 6.19a). In the plane defined by the tilt direction, this angle allows a dense packing of the alkyl segments, compensating for the larger cross-section of the fluorocarbon segment. Hence the tilt plane defines an 'easy' direction along which the monolayer structure can preserve order.

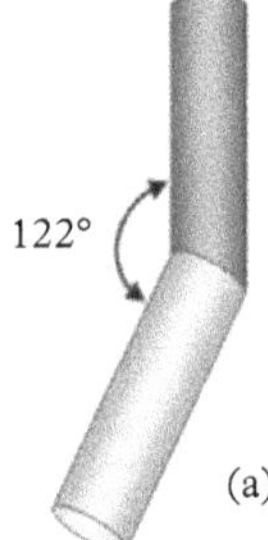
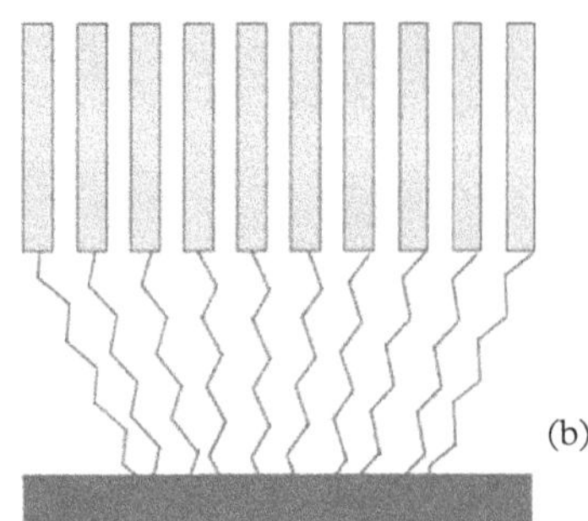

Figure 6.19 *Proposed model for the ribbon structure. (a) Side view of the tilt along the ribbon. (b) View of a cross-section of the ribbon.*

In the direction perpendicular to the tilt plane, the packing is frustrated. A cartoon model of a cross-section of a ribbon is given in Figure 6.19b (Mourran et al., 2005). The width of the ribbons is assumed to be defined by the maximum support a dense incommensurate packing of hydrocarbons can provide to the close-packed fluorocarbon layer. This maximum defines the finite and regular width of the ribbons.

Finally, we note that along the easy long direction of a ribbon 'kinks' can occur when the tilt of the hydrocarbon support switches from forward to backward. In principle this could provide a mechanism to turn the preferred direction. It is tempting to imagine that in this way spirals can be formed, but there seems to be no reason why the turns at such kinks would rotate in one direction only and not alternate randomly.

6.2.4 Example 2: Lamellar diblock copolymer films

The rich bulk morphology of block copolymers was described in Section 5.2.5. Near a surface the morphology can be dramatically affected by the different affinities of the various blocks for the substrate leading to a difference in the surface free energy. In thin films the local packing of the microdomains—which in the bulk are randomly oriented—can become oriented over large areas. The control and applications of the resulting nanopatterns has developed into a huge field of research with important technological applications (see, for example, Schacher, Rupar, and Manners, 2012). Here, we restrict ourselves to some basic features of lamellar films that are at the basis of these developments. Ordered lamellar films can be simply obtained by spincoating a layer on a substrate (say silicon) and subsequent annealing above the glass transition of the polymers involved. If we assume that one of the blocks has a preference for both interfaces (air and substrate) we arrive at the symmetric situation depicted in Figure 6.20a. Often there will be a difference in polarity between the two blocks, in which case the polar block prefers to be at the substrate. Then we obtain the asymmetric situation of Figure 6.20b.

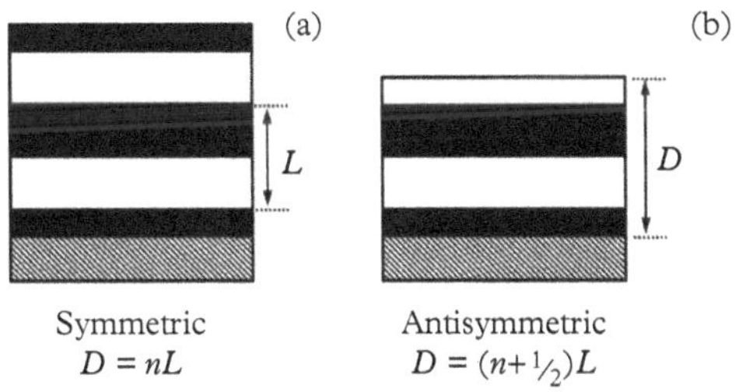

Symmetric
$D = nL$

Antisymmetric
$D = (n+\tfrac{1}{2})L$

Figure 6.20 *Possible lamellar organisation of a diblock copolymer parallel to a substrate.*

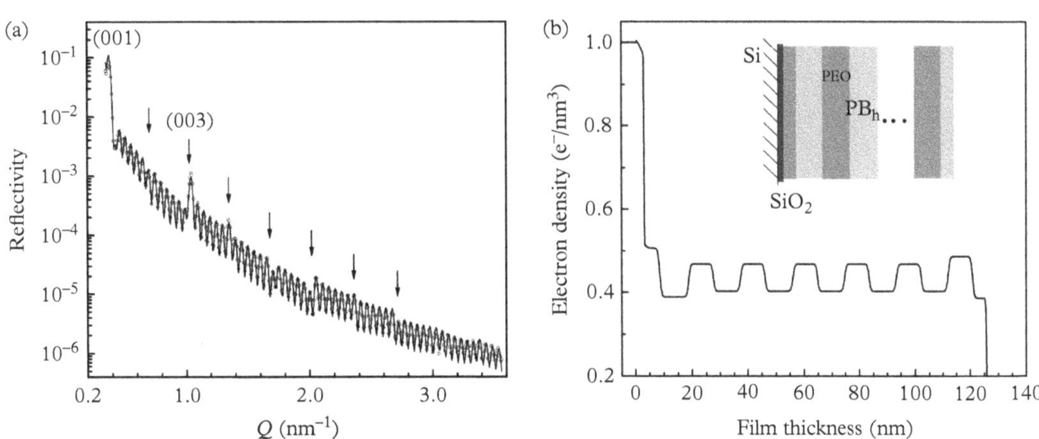

Figure 6.21 *Thin film of lamellar PBh$_{3700}$-b-PEO$_{4300}$ at 90°C. (a) Result for XRR with model fit as full line. (b) Lamellar model; note the sublayer of PEO at the substrate and of PBh at the air interface (after Opitz, Lambreva, and de Jeu, 2002).*

The first (simple) example is PBh$_{3700}$-b-PEO$_{4300}$ of which the bulk SAXS data were given earlier, in Figure 5.23. After spincoating a thin film on a Si-substrate and annealing we expect the polar PEO to go to the substrate and the neutral PBh to prefer the air interface. The XRR results (well above the melting point of PEO) are given in Figure 6.21, together with the model. Many orders of the Bragg-like peaks can be distinguished, though only the first and the third order clearly (compare with Figure 5.23). The lamellar period is given by $L = d_{\text{PEO}} + d_{\text{PBh}} = 8.8 + 9.9 = 18.7$ nm and the total film thickness by $D = 122.9$ nm which is close to 6.5 layers. The contrast is relatively small: the density of PEO is about 15% larger than the density of PBh.

Many polymers do not differ much in their constituting elements. As a result, in block copolymers the difference in (electron) density of the blocks is often small. For example, PS and P2VP have about the same mass density and hardly any contrast in x-ray experiments. In such a case neutron reflectivity is more useful, because it can provide the contrast needed if one of the blocks is deuterated. Let us consider a sample of dPS$_{50k}$-b-P2VP$_{50k}$ in which dPS stands for deuterated PS and $f_{\text{dPS}} = 0.48$ (lamellar morphology). The XRR

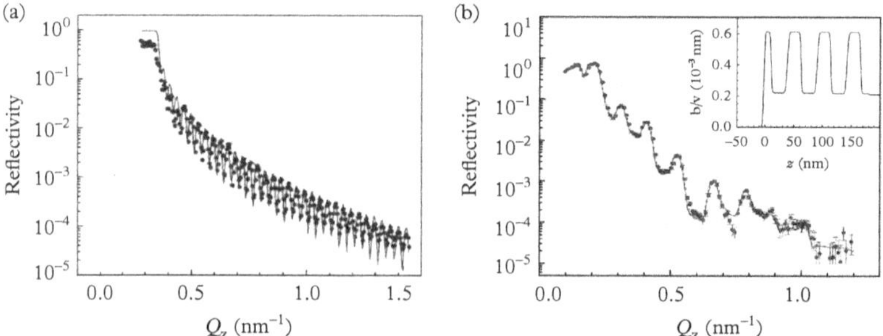

Figure 6.22 *X-ray (a) and neutron reflectivity (b) of a diblock dPS$_{50k}$-b-P2VP$_{50k}$ film, with corresponding fit as full line. Insert: Resulting neutron scattering density profile indicating 3.5 block periods (de Jeu, Lambooy, and Vaknin, 1993).*

results of Figure 6.22a demonstrate the lack of contrast. We only see Kiessig fringes indicating a total film thickness $D = 179$ nm and no further information can be obtained. From a fit to the neutron reflectivity of Figure 6.19b we derive block thicknesses $L_{dPS} = 23.1$ nm and $L_{P2VP} = 28.1$ nm, leading to a block period $L = 51.2$ nm and $D = 3.5 \times L$.

Finally we want to indicate an interesting consequence of the quantized film thickness in terms of the lamellar block periods, as indicated in Figure 6.20. Realizing that the area of the substrate (on which the film is spincoated) also has a specific value, obviously the total volume on the substrate is fixed. Hence, for a certain amount of material, during the annealing process only specific thicknesses are possible. The effect of an ill-matched volume shows up as areas of different film thicknesses (elevations or depressions). Figure 6.23a shows atomic force microscopy of such a situation with about equal contributions from two thicknesses. In the XRR of this sample the difference of one period L between elevations and depressions gives rise to a beat frequency in the reflected intensity; see Figure 6.23b.[11]

[11] Compare to a sound beat of two acoustic signals with a small difference in frequency.

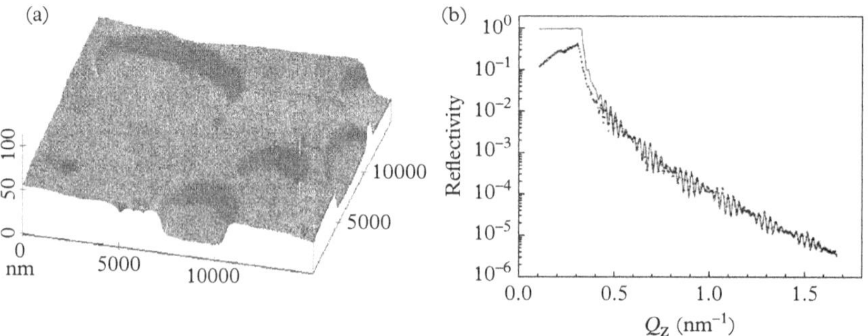

Figure 6.23 *Quantized structure of a lamellar block copolymer film. (a) AFM picture with flat elevations and depressions giving a height difference of 29.4 nm. (b) XRR with a fit corresponding to an addition of two independent models with $D = 193.5/222.9$ nm $= 6.5/7.5 \times 29.7$ nm (after de Jeu et al., 1993).*

6.3   Grazing incidence diffraction

So far we have discussed reflected x-rays above the critical angle of total reflection $\theta_c \equiv \alpha_c$ (see, once more, Figure 6.1) which allows us to obtain structural information along the surface normal. To obtain lateral information (along the surface) we consider the situation below α_c, as depicted in Figure 6.24a. It should be realized that total reflection cannot happen by itself and that some penetration must occur into the reflecting material (full line in Figure 6.24b). This involves the presence of an evanescent wave parallel to the interface, the intensity of which decays exponentially along the normal. This evanescent wave can be considered as a secondary source of x-rays to be used for in-plane scattering. Thus we can obtain structural information of the upper sample region close to the interface. This situation is sketched in Figure 6.25a, where the in-plane scattering angle is indicated as ψ. Scanning this angle using a linear detector perpendicular to the sample plane, a two-dimensional picture of the surface scattering can be obtained: GID, grazing-incidence diffraction, or GIWAXS.[12] If the scattering angle is small (GISAXS) a beamstop is needed and usually a two-dimensional detector is used (see Figure 6.25b).[13] As with conventional diffraction, the difference between GIWAXS and GISAXS is not fundamental, but more of a practical nature: GIWAXS for in-plane structures < 1 nm, typically atomic length scales, GISAXS for larger assemblies.[14]

Figure 6.24b still deserves a few comments. Obviously, without absorption the penetration depth Λ above the critical angle is infinite (transmission) and no evanescent wave can be present. In practice, Λ will still be finite and will depend on the degree of absorption. Hence GID is also possible above the critical angle and in all cases the penetration depth (and thus the relevant surface slab

[12] As GIWAXS is older than GISAXS, in the early literature GID is often used for GIWAXS. To avoid confusion we shall exclusively use GIWAXS next to GISAXS.

[13] Compare with the situation in Figure 5.1 for conventional SAXS.

[14] A nice tutorial of GISAXS and GIWAXS with many examples is given by Detlef M. Smilgies (CHESS synchrotron, Cornell, USA) and can be found at: staff.chess.cornell.edu/~smilgies/gisaxs/GISAXS.php.

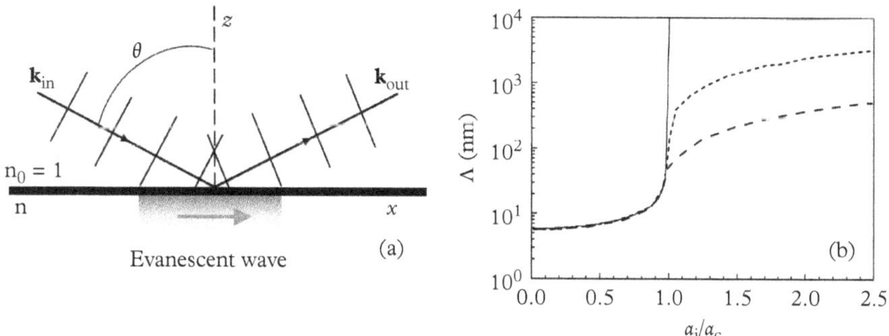

Figure 6.24 *(a) Schematic picture of the evanescent wave that penetrates the sample at specular reflection. (b) Penetration depth (1/e value) in Si as a function of the incident angle for various values of the absorption coefficient. From top to bottom: β/δ = 0, 0.05, and 0.1.*

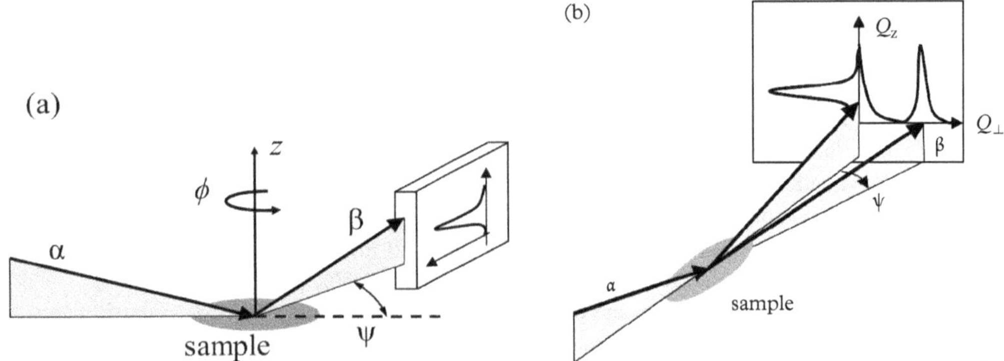

Figure 6.25 *Configurations for grazing-incidence diffraction at wide angles (a, GIWAXS) and at small angles (b, GISAXS).*

investigated) can be varied by choosing the incident angle. In the kinematical theory as employed so far in SAXS data analysis, the magnitude of the x-ray electric field does not change over the x-ray path and multiple scattering is ignored (Born approximation). In grazing-incidence scattering, the electric field intensity normal to the surface is redistributed. This leads to: (a) an incident wave amplitude that varies at different height, and (b) a much higher chance that the scattered beam will be re-scattered again. Hence the theoretical interpretation of grazing-incidence scattering is more elaborate. The Distorted Wave Born approximation (DWBA) has been developed to account for these dynamical phenomena. For the purpose of this book, we do not go into these aspects, but restrict the remaining part of this section to demonstrate, using two examples, the power of combining x-ray reflectivity and grazing-incidence scattering to study soft-matter films and surfaces.

6.3.1 Surface freezing of alkyl monolayers

The properties of the series of alkanes C_nH_{2n+2} are interesting in various respects. First, they provide examples of so-called solid-rotator phases: crystals with a strong rotational freedom of the molecules. In addition, the longer ones can be regarded as precursors for polymers, especially regarding the study of crystallization (compare with Section 4.4). During the last decade it has been found that for a certain range of length n, they can display a very specific surface property: surface freezing. Upon cooling from the liquid state a crystalline monolayer develops on the surface of the bulk liquid, at a temperature a few degrees above its freezing temperature. Upon cooling no additional layers are formed until full crystallization is reached. The ordered monolayer phase appears for $16 \leq n \leq 50$.

The x-ray evidence for this peculiar behaviour is displayed in Figure 6.26a (XRR) and Figure 6.26b (GIWAXS). At high temperatures XRR shows the usual Fresnel decay of a liquid surface (compare with Figure 6.7). Upon approaching the crystallization temperature in XRR, broad Kiessig fringes appear indicating a monolayer on top. The proposed structure is shown in Figure 6.27. Figure 6.26b shows GIWAXS of the C20 alkane liquid. At high temperatures only a broad bulk liquid peak is observed. Formation of the crystalline monolayer is manifested by a drop in intensity of the liquid peak and the appearance of a narrow Bragg peak (solid circles). The peak position is consistent with that of a solid rotator phase. The molecules are oriented on average normal to the layer and packed hexagonally.

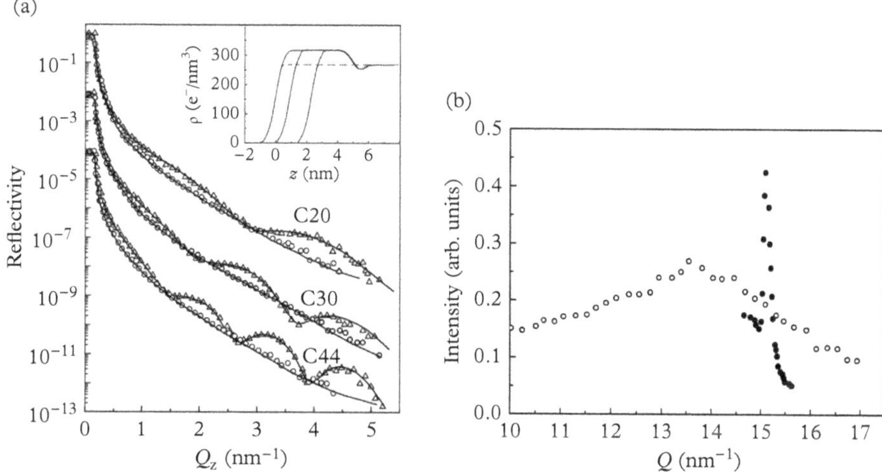

Figure 6.26 *(a) XRR of alkanes with n = 20, 30, and 44. Circles: data at high temperatures; triangles: data indicating surface freezing at lower temperatures. (b) In-plane GIWAXS of the C20 alkane. Open circles: broad bulk liquid peak. Solid circles: narrow Bragg peak that is observed at lower temperatures parallel to the appearance of Kiessig fringes. (Reprinted with permission from Wu et al., 1993. Copyright American Physical Society.)*

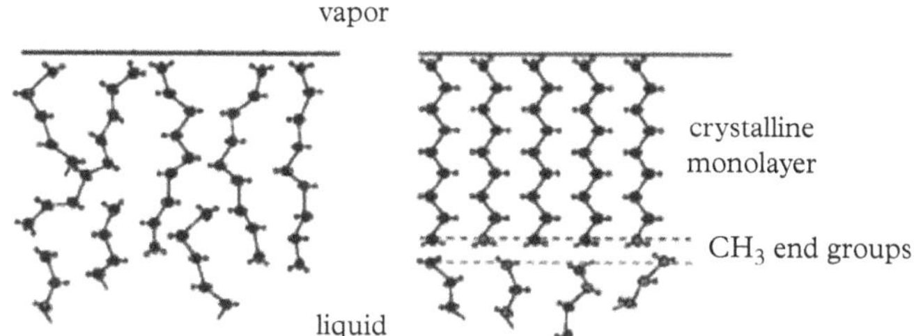

Figure 6.27 *Model for the two situations depicted in Figure 6.26.*

6.3.2   Controlled orientations in a smectic block copolymer film

For a second example we return to the smectic-amorphous block copolymer P(MMA-F8H2A) with fluorinated side chains, seen in Figure 5.34. In the bulk the smectic layers are oriented perpendicular to the block interfaces, which poses the interesting question about the orientation in a thin film. Figure 6.28a shows XRR[15] of a film of the compound F2 that has a lamellar-*within*-hexagonal structure. The two main characteristics are the Kiessig fringes from the total film thickness and a Bragg peak at $Q = 2.00\,\mathrm{nm}^{-1}$ corresponding to a period of 3.14 nm. The best fit to the XRR data, given as the continuous line in Figure 6.28a, indicates a density profile composed of 21 smectic layers with a period of 3.2 nm. This regular part is sandwiched between a monolayer adjacent to the substrate of 1.4 nm thickness and a top layer of 10.3 nm with a considerable roughness. The periodicity calculated from the Bragg peak is very close to the bulk smectic spacing of 3.3 nm. This indicates that the smectic layers are aligned parallel to the substrate with the side chains anchoring to the film interfaces.

Figure 6.28b shows a lateral GISAXS peak at 30.4 nm, close to the domain period in the bulk (30 nm). Hence, the cylindrical domains are oriented normal to the substrate. A model incorporating these observations is shown in Figure 6.29. This peak can be well fitted by a Lorentzian function leading to a correlation length of 40 nm. This relatively small value indicates absence of any lateral long-range order of the upright cylindrical domains. The GIWAXS result of Figure 6.28c shows a peak at $Q = 12.44\,\mathrm{nm}^{-1}$. This corresponds to a spacing of 0.5 nm, which is consistent with the interchain distance of the side chains in the CrB phase, in agreement with the bulk observations.

In conclusion, in thin films of the cylinder-forming block copolymer F2, the smectic layering and the orientational wetting properties of the fluorinated side chains stabilize the minority cylindrical domains normal to the substrate.

[15] Note that the reflected intensity has been divided by the Fresnel reflectivity R_F of the Si wafer ($Q_c = 0.32\,\mathrm{nm}^{-1}$) to bring the relevant features clearly out.

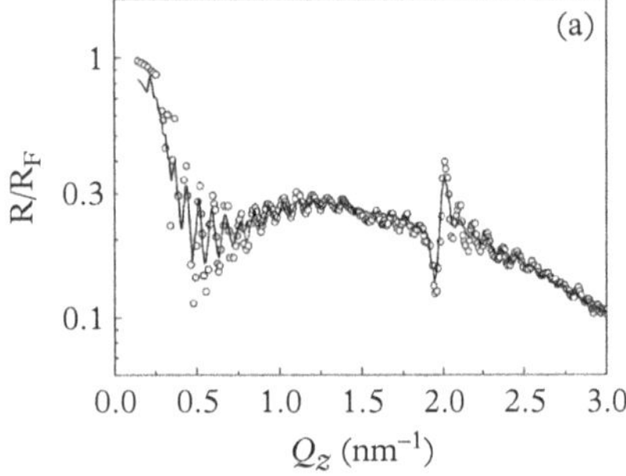
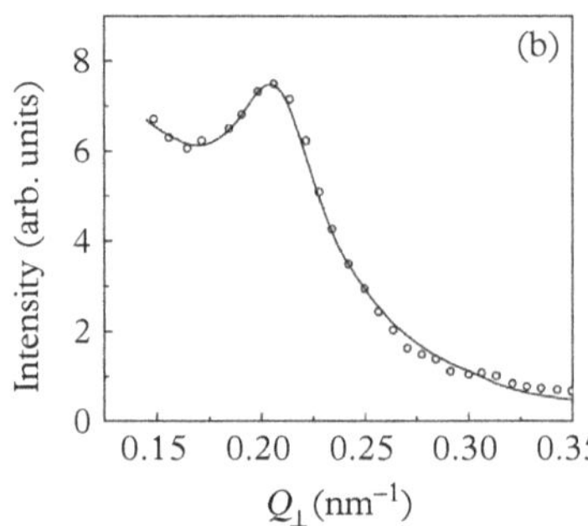
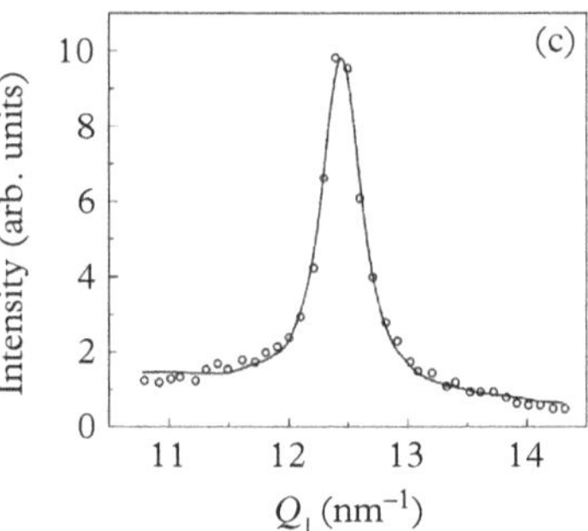

Figure 6.28 *Experimental x-ray result for a thin film of the F2 system. (a) XRR. (b) GISAXS with a Lorentzian fit. (c) GIWAXS (after Al-Hussein et al., 2005).*

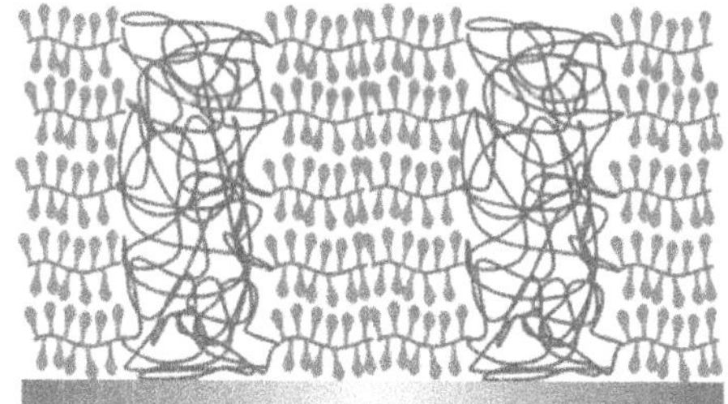

Figure 6.29 *Model of the thin-film structure of compound F2 from the data of Figure 6.28.*

A model of the structure as shown in Figure 6.29 can be nicely derived by normal-incidence (XRR) and grazing-incidence (GISAXS and GIWAXS) x-ray scattering. The absence of long-range order of the cylinders has been confirmed by atomic force microscopy.

6.4 Case study: Monolayer structure of substituted thiophenes

Conjugated organic materials are interesting for developments in the field of organic semi-conductors (organic field-effect transistor, OFET). These transistors are the main logic units in electronic circuits, in which they usually function as either a switch or an amplifier. In this context, oligothiophenes of the structure R-nT-R′ are remarkable materials. Here nT stands for a series of conjugated thiophene rings with typically $n = 4$–7 while R and R′ are substituents. These compounds possess a relatively large electric mobility thanks to transverse π–π coupling between neighbouring oligothiophene cores. A typical crystal structure of unsubstituted 6T is shown in Figure 6.30. The molecular long axis is at an angle of 23.5° with the a-axis, and the molecule is practically planar. The unit cell contains four molecules in a herringbone packing, common to most planar molecules.

In an organic thin-film transistor the carrier transport in the channel between the source and the drain electrodes occurs in the first few layers of the semiconductor. Hence, the structure and properties of ultrathin films have received special attention. Conduction in thin polycrystalline films is controlled by the degree of crystallinity and the extent of the crystalline domains, domain boundaries, crystal defects, and impurities. Optimizing single-crystal devices in monolayers or thin multilayers involves the delicate point of decoupling of lateral (two-dimensional) crystal growth from that perpendicular to the substrate, leading to multilayers. A variety of complex phenomena plays a role, including absorption and dewetting processes and surface freezing. The growth kinetics of molecular thin films is influenced by the anisotropy in molecular structure and crystal packing in relation to the interaction with the substrate. In the case of oligothiophenes, variation of the substituents R and R' has been used to control the anisotropy (including

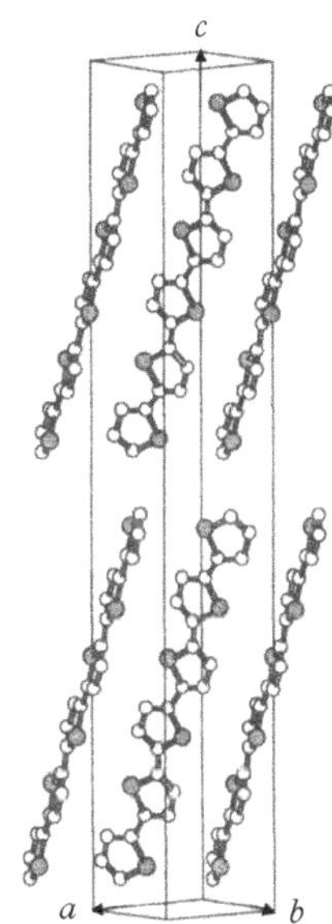

Figure 6.30 *Monoclinic crystal structure of 6T, with a large dimension c = 4.47 nm, a = 0.785 nm, and b = 0.603 nm. The monoclinic angle is close to 90°; hence deviations from an orthorombic structure are small. (Reprinted with permission from Horowitz et al., 1995. Copyright American Chemical Society.)*

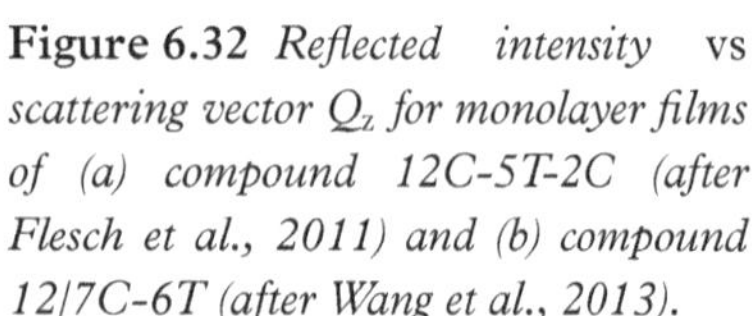
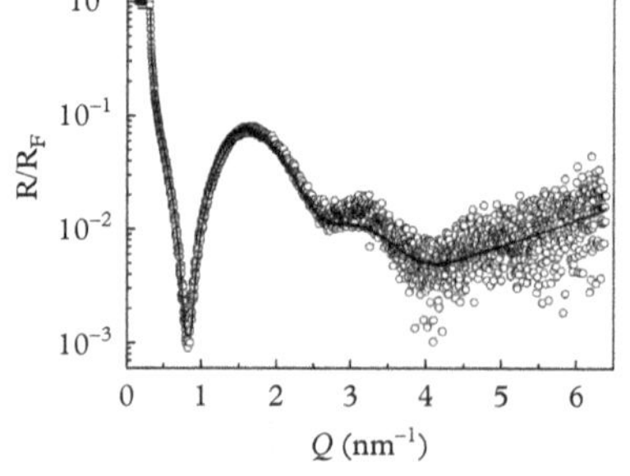
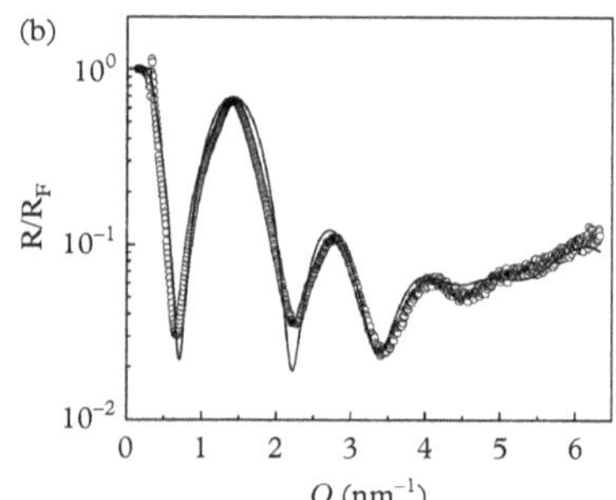

Figure 6.31 *Structure of the substituted thiophenes discussed and the abbreviations used for the compounds.*

possible liquid-crystalline order). Some examples of such substituted structures are displayed in Figure 6.31. Obviously for the study of thin-layer systems, XRR and GIWAXS are valuable tools.

We shall first concentrate on the upper two compounds of Figure 6.31. The first one, 12C-5T-2C, exhibits a single melting temperature of 247°C, compound 12/7C-6T melts at 232°C (considerably lower than unsubstituted 6T). The end group of 12C-5T-2C allows the compound to be chemisorbed on a silicon substrate; compound 12/7C-6T will be physisorbed after spincoating. An important aspect of packing these molecules inside a monolayer is the difference between the Van der Waals diameters of $d_{\mathrm{alkyl}} \approx 4$–5 nm for a compact alkyl chain compared to $d_{\mathrm{thio}} \approx 3$–4 nm for the oligothiophene core. From these numbers one could expect the first compound to be standing approximately upright at the substrate (optimum π–π interactions); for the latter one this might be difficult because of the branched alkyl chain. The sterically demanding end groups will hinder a parallel orientation of the thiophene cores. Tilting of the central cores with respect to the end groups is one possibility to solve this problem.

Figure 6.32 shows XRR data for both compounds with fits according to the modelling given in Table 6.2. The result for 12C-5T-2C can be described by two sublayers: an electron-rich thiophene part and a less dense alkyl chain attached

Figure 6.32 *Reflected intensity vs scattering vector Q_z for monolayer films of (a) compound 12C-5T-2C (after Flesch et al., 2011) and (b) compound 12/7C-6T (after Wang et al., 2013).*

Table 6.2 *Fitting results corresponding to the XRR of Figure 6.32.*

		12C-5T-2C	12/7C-6T
Top layer	thickness	-	0.96 nm
	electron density		240 e/nm³
	roughness		0.24 nm
Middle layer	thickness	2.06 nm	2.4 nm
	electron density	370 e/nm³	460 e/nm³
	roughness	0.7 nm	2.4 nm
bottom layer	thickness	1.56 nm	1.25 nm
	electron density	250 e/nm³	360 e/nm³
	roughness	0.38 nm	0.38 nm

to the substrate. Modelling films of 12/7C-6T requires a sandwich structure with three sublayers: a middle thiophene layer with the highest electron density between two layers of alkyl substituents with a lower electron density at the silicon/film and the film/air interface. The thicknesses of the latter two sublayers are close. This suggests similar surface structures, presumably consisting of the branched alkyl group. As the outer-layer thickness is substantially smaller than the length of the alkyl chain (about 1.5 nm), there must be either considerable disorder or a strong tilt. The thickness of the middle sublayer (2.4 nm) is comparable to the length of a 6T molecule (about 2.2 nm).

More information can be obtained by GIWAXS, shown for each of the two samples in Figure 6.33. The first one shows Bragg rods that extend along the vertical z-axis, though only over a limited range. They are found at $Q_\perp = 13.9\,\mathrm{nm}^{-1}$, $16.1\,\mathrm{nm}^{-1}$, and $19.7\,\mathrm{nm}^{-1}$, leading to a two-dimensional unit cell with lattice constants $a = 0.78\,\mathrm{nm}$ and $b = 0.554\,\mathrm{nm}$ (Mathijssen et al., 2009). These values are

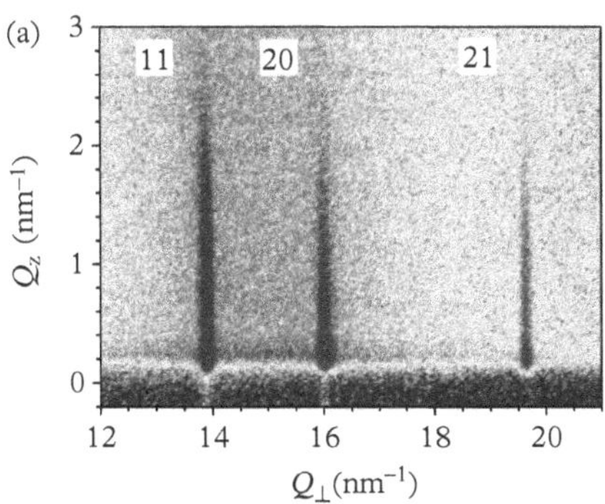

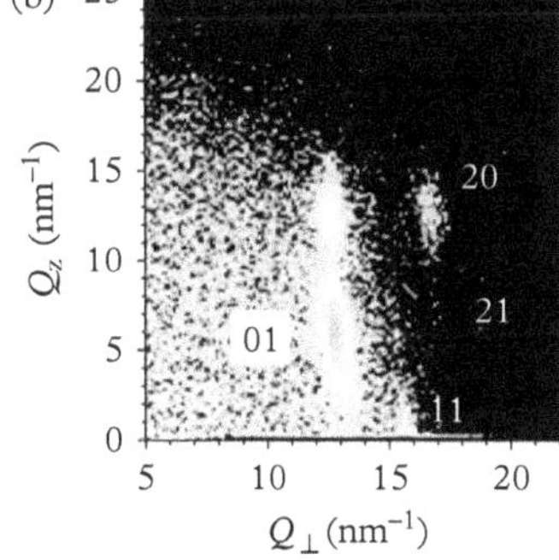

Figure 6.33 *GIWAXS of (a) a monolayer of 12C-5T-2C [reprinted with permission from Flesch et al. (2011), copyright American Chemical Society] and (b) 12/7C-6T (after Wang et al. 2013), respectively, with the assignment (hk) of the two-dimensional lattice indicated.*

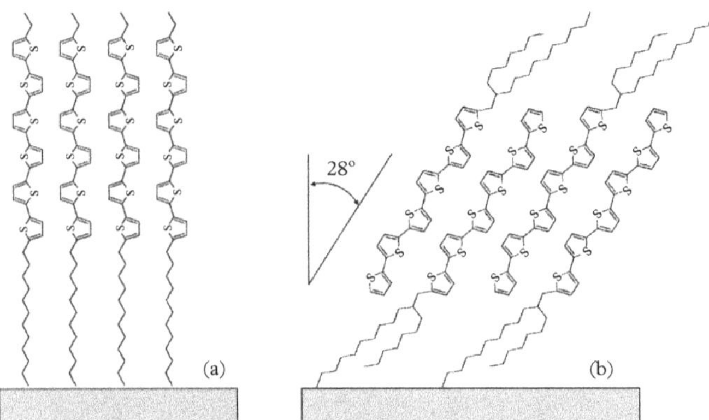

Figure 6.34 *Two-dimensional picture of the proposed packing of (a) compound 12C-5T-2C, and (b) for compound 12/7C-6T. The anticipated herringbone character needs a third dimension.*

close to the ones of sexithiophene (Figure 6.30). The perpendicular distance between two neighbouring sexithiophene molecules is close to the a-value. Hence, the lateral packing of the thiophene units in 12C-5T-2C must be responsible for the two-dimensional crystals and these units pack upright in a herringbone pattern (see Figure 6.34a). Flesch et al. (2011) also monitored the changes in the GIWAXS pattern upon heating. After annealing at 127°C, a transition occurred from upright-standing molecules (Bragg rods of Figure 6.33a) to a phase of tilted molecules. The unit cell remained rectangular, as in the original sample. The projection of the long molecular axis onto the ab-plane was along the b-axis. Hence, the unit-cell parameter b increased while a remained constant upon tilting of the molecules.

The situation for 12/7C-6T is somewhat more complicated (see Figure 6.33b and Table 6.3). In Figure 6.34b a model is presented in which the outer sublayers consist of the bulky branched alkyl chains oriented alternatively up and down, adhered to a central thiophene sublayer that, in turn, is stabilized by π–π interactions. The necessary space required to pack these chains is obtained by

Table 6.3 *Assignment of the x-ray peaks of 12/7C-6T (Figure 6.33b).*

(hk)	$Q_\perp$ (nm^{-1}) experimental	$Q_\perp$ (nm^{-1}) calculated	Q_z (nm^{-1}) experimental
01(1)	12.6	12.6	5.1
01(2)	12.6	12.6	10.3
20	16.6	16.5	10.4
11	15.6	15.0	0
21	18.7	20.8	5.3

Table 6.4 *In-plane lattice constants of 6C-6T-6C for various thicknesses.*

	b (nm)	c (nm)
Monolayer	0.89	0.63
Multilayer	0.84	0.59
Bulk	0.79	0.58

interdigitation of the central 6T core. Some tilt must be involved in order to accommodate the differences in cross section of the different groups.

The general features of the GIWAXS picture can be understood in analogy to the interpretation of scattering data of hexagonal Langmuir monolayers of fatty acids by Kaganer et al. (1999). Using the model of Figure 6.34b we assume a centred rectangular packing with two molecules per unit cell. With these assumptions we calculate[16] a tilt angle of 28° and lattice vectors $a = 0.76$ nm and $b = 0.50$ nm.[17] This leads to an area per molecule of $ab/2 = 0.19$ nm^2. Using $n = 252$, the number of electrons of 6T, and the thickness of the 6T middle layer from XRR (2.4 nm, see Table 6.2), we calculate an electron density of 553 e$^-$/nm^3. Projected on the layer normal this gives 488 e/nm^3, in good agreement with the electron density of 460 e$^-$/nm^3 from XRR.

Mannebach et al. (2013) investigated mono- and multilayers of 6C-6T-6C. The linear alkyl chains do not prevent optimum π–π packing of the thiophene units in an upright orientation with respect to the substrate. This is similar as that discussed above for the asymmetric substituents of 12C-5T-2C. Table 6.4 gives results for the two-dimensional lattice parameters for various film thicknesses. The lattice of the monolayer is expanded with respect to the multilayer, which is in turn is larger than the bulk values. Evidently, restricting the third dimension leads to more stringent in-plane conditions.

The study of (mono)layers of substituted thiophenes is a broad field with fundamental and practical interest. In the examples shown we find—not surprisingly for a monolayer—a smectic-like packing for which the lateral π–π coupling of the thiophene units is mainly responsible. However, combining elongated molecular units in a parallel way requires a certain matching of their transverse dimensions to reduce packing frustration. As we have seen, this can be reached by either tilting one unit with respect to the other or by (partial) overlapping (interdigitation). In such a situation a fundamental difference arises between bulk and thin-film structures: Neighbouring overlapping structures in bulk have no strict equivalent in a monolayer. In the example of Figure 6.34b this was solved by the specific construction of a interdigitated bilayer as first single layer. In addition, of course, particular surface interactions may play a role in monolayer formation.

[16] In fact the tilt angle is given by $\tan\theta = Q_z^{11}/\sqrt{(Q_\perp^{11})^2 - (Q_z^{11}/2)^2}$ (Kaganer, Möhwald, and Dutta, 1999).

[17] These numbers follow directly from the $Q_\perp$-values of the (20) and (01) reflections in Figure 6.33b and Table 6.3.

Bibliography

General references

Als-Nielsen J, McMorrow D (2011) *Elements of Modern X-Ray Physics*. Second edition. Wiley. [Modern complete physical text]

Hukins DWL (1981) *X-ray Diffraction by Disordered and Ordered Systems*. Pergamon. [High-quality accessible short text]

Jones RAL (2002) *Soft Condensed Matter*. Oxford University Press. [Introduction to the physics of soft matter]

Roe R-J (2000) *Methods of X-Ray and Neutron Scattering in Polymer Science*. Oxford University Press. [Reference text for scattering of polymers]

Schnablegger H, Sing Y (2003) *The SAXS Guide*. Second edition. Anton Paar. [Useful practical only guide]

Sivia DS (2011) *Elementary Scattering Theory*. Oxford University Press. [Accessible discussion of the mathematics of scattering]

Tolan M (1999) *X-Ray Scattering from Soft Matter Thin Films*. Springer. [Accessible text for scattering from interfaces and films]

Warren BE (1990) *X-Ray Diffraction*. Dover. [Classical physical reference text]

Useful links

http://henke.lbl.gov/optical_constants (*x-ray properties of materials*)

http://mathworld.wolfram.com (*mathematics*)

http://www.xtal.iqfr.csic.es/Cristalografia/index-en.html (*general course in crystallography*)

References cited in the text

Al-Hussein M, Séréro Y, Konovalov O, Mourran A, Möller M, de Jeu WH (2005) *Macromol* 38:9610

Bailey TS, Pham HD, Bates FS (2001) *Macromol* 34:6994

Bates FS (1991) *Science* 251:898

Bates FS, Frederickson GH (1999) *Physics Today* 52:32

Billard J, Dubois JC, Vaucher C, Levelut AM (1981) *Mol Cryst Liq Cryst* 66:115

Bouwman WG, de Jeu WH (1993) In A Buka (ed) *Modern Topics in Liquid Crystals*. Singapore: World Scientific, p. 161

Bouwman WG, de Jeu WH (1994) *J Physique II* 4:787

Bragg WL (1965) *The Crystalline State. A General Survey* (Cornell)

Braslau A, Deutsch M, Pershan PS, Weiss AH, Als-Nielsen J, Bohr J (1985) *Phys Rev Lett* 54: 114

Brock JD, Birgeneau RJ, Litster JD, Aharony A (1989) *Contemp Physics* 30:321

Bunn CW (1939) *Trans Faraday Soc* 35:482

van der Burgt FPTJ, Rastogi S, Chadwick JC, Rieger B (2002) *J Macromol Sci Phys B* 41:1091

Byelov D, Panine P, Remerie K, Biemond E, Alfonso GC, de Jeu WH (2008) *Polymer* 49:3076

Chandrasekhar S, Prasad SK, Rao DSS, Balagurusamy VSK (2002) *Proc Indian Nat Sci Acad* A 68:175

Chao C-Y, Chou C-F, Ho JT, Hui SW, Jin A, Huang CC (1996) *Phys Rev Lett* 77:2750

Daillant J, Gibaud A (2009) (eds) X-*ray and Neutron Reflectivity*. Berlin: Springer.

Danauskas SM, Li D, Meron M, Lin B, Lee KYC (2008) *J Appl Cryst* 41:1187

Davidson P, Gabriel JCP (2005) *Curr Opin Colloid Interface Sci* 9:377

Fera A, Ostrovskii BI, Sentenac D, Samoilenko I, de Jeu WH (1999) *Phys Rev E* 60:5033

Flesch HG, Mathijssen SGJ, Gholamrezaie F, Moser A, Neuhold A, Novák J, Ponomarenko SA, Shen Q, Teichert C, Hlawacek G, Puschnig P, Ambrosch-Draxl C, Resel R, de Leeuw DM (2011) *J Phys Chem C* 115:22925

Fontes E, Heiney PA, de Jeu WH (1988) *Phys Rev Lett* 61:1202

Förster S, Konrad M (2003) *J Mater Chem* (2003) 13: 2671

Glatter O (2002) In P Lindner, T Zemb (eds) *Neutrons, X-Rays and Light: Scattering Methods Applied to Soft Condensed Matter*. Amsterdam: Elsevier, p. 73

Gramsbergen EF, de Jeu WH, Als-Nielsen J (1986) *J Phys France* 47:711

Horowitz G, Bachet B, Yassar A, Lang P, Demanze F, Fave J-L, Garnier F (1995) *Chem Mater* 7:1337

Ikkala O, Houbenov N, Rannou P (2014) In JW Goodby, PJ Collings, T Kato, C Tschierske, HF Gleeson, P Raynes (eds) *Handbook of Liquid Crystals*. Vol 7, second edition. Wiley-VCH, p. 541

Jenkins R, Snyder R. (1996) *Introduction to X-Ray Powder Diffractometry*. Wiley.

de Jeu WH, Lambooy P, Hamley IW, Vaknin D, Pedersen JK, Kjaer K, Seyger R, van Hutten P, Hadzioannou G (1993) *J Phys II France* 3:139

de Jeu WH, Lambooy P, Vaknin D (1993) *Macromol* 26:4973

de Jeu WH, Ostrovskii BI (2012) *Adv Polym Sci* 250: 187

de Jeu WH, Ostrovskii BI, Shalaginov AN (2003) *Rev Mod Phys* 75:181

de Jeu WH, de Poorter, JA (1977) *Phys Lett* 61A:114

Kaganer VM, Möhwald H, Dutta P (1999) *Rev Mod Phys* 71:779

Keller A (1968) *Rep Prog Phys* 31:623

Khandpur AK, Forster S, Bates FS, Hamley IW, Ryan AJ, Bras W, Almdal K, Mortensen K (1995) *Macromol* 28:8796

Koltover I, Salditt T, Rädler JO, Safinya CR (1998) *Science* 281:78

Kornfield JA, Kumaraswamy G, Issaian AM (2002) *Ind Eng Chem Res* 41:6383

Kraack H, Deutsch M, Sirota EB (2000) *Macromol* 33:6174

Krafft MP, Riess JG (2009) *Chem Rev* 109:1714

Kroon M, Vos WL, Wegdam GH (1998) *Phys Rev E* 57:1962

Kuijk A, van Blaaderen A, Imhof A (2011) *J Am Chem Soc* 133:2346

Lambreva DM (2005) *Thesis*, Eindhoven University of Technology

Lambreva DM, Opitz R, Reiter R, Frederik PM, de Jeu WH (2005) *Polymer* 46:4868

Leadbetter AJ (1979) In GR Luckhurst, GW Gray (eds) *The Molecular Physics of Liquid Crystals*. London: Academic Press, p. 285

Leadbetter AJ, Frost JC, Mazid MA (1979) *J Physique Lett* 40:325

van der Lee A, Salah F, Harzallah B (2007) *J Appl Cryst* 40:820

Lemaire BJ, Davidson P, Ferré J, Jamet JP, Panine P, Dozov I, Jolivet JP (2002) *Phys Rev Lett* 88:125507

Li L, Lambreva DM, de Jeu WH (2004) *J Macromol Sci B Physics* 43:59

Loo Y-L, Register RA, Ryan AJ (2002) *Macromol* 35:2365

Mai S-M, Fairclough JPA, Hamley IW, Matsen MW, Denny RC, Liao B-X, Booth C, Ryan AJ (1996) *Macromol* 29:6212

Mannebach EM, Spalenka JW, Johnson PS, Cai Z, Himpsel FJ, Evans PG (2013) *Adv Funct Mater* 23:554

Mathijssen SGJ, Smits ECP, van Hall PA, Wondergem HJ, Ponomarenko SA, Moser A, Resel R, Bobbert PA, Kemerink M, Janssen RAJ, de Leeuw DM (2009) *Nature Nano* 4:674

Megens M, van Kats CM, Bösecke P, Vos WL (1997) *Langmuir* 13:6120

Mezzenga, R, Meyer C, Servais C, Romoscanu AI, Sagalowicz L, Hayward RC (2005) *Langmuir* 21:3322

Mol EAL, Wong GCL, Petit J-M, Rieutord F, de Jeu WH (1998) *Physica B* 248:191

Mourad MCD, Byelov DV, Petukhov AV, de Winter DAM, Verkleij AJ, Lekkerkerker HNW (2009) *J Phys Chem B* 113:11604

Mourran A, Tartsch B, Gallyamov M, Magonov S, Lambreva D, Ostrovskii BI, Dolbnya IP, de Jeu WH, Möller M (2005) *Langmuir* 21:2308

Natta G, Corradini P (1960) *Il Nuovo Cimento* 15:40

Obraztsov EP, Muresan AS, Ostrovskii BI, de Jeu WH (2008) *Phys Rev E* 77:021706

Opitz R, Lambreva DM, de Jeu WH (2002) *Macromol* 35:6930

Ostrovskii BI, Sulyanov SN, Boiko NA, Shibaev VP, Astaf'ev SB, Yanusova LG, de Jeu WH (2013) *Eur Phys J E* 36:134

Panine P, Di Cola E, Sztucki N, Narayanan T (2008) *Polymer* 49:676

Pedersen JS (1997) *Adv Colloid Interface Sci* 70:171

Peiser HS, Rooksbym HP, Wilson AJC (1955) *X-ray diffraction by polycrystalline materials*. London: The Institute of Physics

Penfold J (2001) *Rep Prog Phys* 64:777

van den Pol E, Thies-Weesie DME, Petukhov AV, Panine P, Vroege GJ (2008) *J Phys Condens Matter* 20:404219

Pynn R (1990) *Los Alamos Science* 19:1

Rädler JO, Koltover I, Salditt T, Safinya CR (1997) *Science* 275:810

Rastogi S, Hikosaka M, Kawabata H, Keller A (1991) *Macromol* 24:6384

Rastogi S, Kurelec L, Lemstra PJ (1998) *Macromol* 31:5022

Richardson RM, Agina EV, Boiko NI, Shibaev VP, Grillo I (2008) *J Phys Chem B* 112:16346

Schacher FH, Rupar PA, Manners I (2012) *Angew Chem Int Ed* 51:7898

Safinya CR. (2001) *Curr Opin Struct Biology* 11:440

Schmid A (2014) *Private communication*

Seddon JM (1990) *Biochimica Biophysica Acta* 1031:1

Seddon JM, Robins J, Gulik-Krzywicki T, Delacroix H (2000) *Phys Chem Chem Phys* 2:4485

Skar-Gislinge N, Arleth L (2011) *Phys Chem Chem Phys* 13:3161

Skar-Gislinge N, Simonson JB, Mortensen K, Feidenhans'l R, Sligar SG, Møller BL, Bjørnholm T, Arleth L (2010) *J Am Chem Soc* 132:13713

Smits ECP, Mathijssen SGJ, van Hal PA, Setayesh S, Geuns TCT, Mutsaers KAHA, Cantatore E, Wondergem HJ, Werzer O, Resel R, Kemerink M, Kirchmeyer S, Muzafarov AM, Ponomarenko SA, de Boer B, Blom PWM, de Leeuw DM (2008) *Nature* 455: 956

Soper AK, Bruni F, Ricci MA (1997) *J Chem Phys* 106:247

Strobl GR (2006) *Prog Polym Sci* 31:398

Strobl GR (2007) *The Physics of Polymers*. Berlin: Springer

Strobl GR (2009) *Rev Mod Phys* 81:1287

Svergun DI (2007) *J Appl Cryst* 40:s10

Tresset G (2009) *PMC Biophysics* 2:3

Turner-Jones A, Aizlewood JM, Beckett DR (1964) *Makromol Chem* 75:134

Ungar G, Stejny J, Keller A, Bidd I, Whiting MC (1985) *Science* 229:386

Vlieg E, de Vries SA, Alvaraz J, Ferrer J (1997) *J Synchrotron Rad* 4:210

Wang J, de Jeu WH, Ziener U, Polinskaya MS, Ponomarenko SA, Rücker U, Ruderer MA, Herzig EM, Müller-Buschbaum P, Möller M, Mourran A (2013) *Langmuir* 30:27527

Wu XZ, Sirota EB, Sinha SK, Ocko BM, Deutsch M (1993) *Phys Rev Lett* 70:958

Yarnell JL, Katz MJ, Wenzel RG, Koenig H (1973) *Phys Rev A* 7:2130

Yethniraj A, van Blaaderen A (2003) *Nature* 421:513

The manufacturer's authorised representative in the EU for product safety is
Oxford University Press España S.A. of el Parque Empresarial San Fernando de
Henares, Avenida de Castilla, 2 – 28830 Madrid (www.oup.es/en or product.
safety@oup.com). OUP España S.A. also acts as importer into Spain of products
made by the manufacturer.